Conway R Howard

Earthwork Mensuration

Conway R Howard

Earthwork Mensuration

ISBN/EAN: 9783337398835

Printed in Europe, USA, Canada, Australia, Japan

Cover: Foto ©berggeist007 / pixelio.de

More available books at **www.hansebooks.com**

EARTHWORK MENSURATION,

ON THE BASIS OF THE

PRISMOIDAL FORMULA.

Containing a Simple and Labor-saving Method of
OBTAINING PRISMOIDAL CONTENTS DIRECTLY FROM END AREAS.

ILLUSTRATED BY EXAMPLES,
AND ACCOMPANIED BY PLAIN RULES FOR PRACTICAL USE.

BY
CONWAY R. HOWARD,
CIVIL ENGINEER, RICHMOND, VA.

NEW YORK:
D. VAN NOSTRAND, PUBLISHER,
23 MURRAY AND 27 WARREN STREET.
—
1874.

PREFACE.

THIS work claims to present a new and systematized method of finding the prismoidal contents of Earthwork by means of Tables accompanied by Rules so plain and simple of application as to fit it for the common uses of Engineers.

When the ratios of the side slopes are constant between end sections of which the transverse surface lines are sensibly similar, all ordinary cases of thorough cut and fill, terminal pyramids, side-hill work, and borrow pits are covered by Formulæ (17), (18), and (19), and the prismoidal contents for all side slopes and bases are taken from Tables 4 and 5 by Rules (1), (2), and (3).

In the method used, the heights of equivalent level sections are not involved, nor is any calculation needed for 100-feet lengths beyond ascertaining the half-sum and the difference of two quantities. For the most part Tables do the work of the calculator, and any one who can approximate cubic contents by the rough method of "Average Areas" is competent to obtain the prismoidal contents by the Rules given.

The tables of level cuttings are not needed when areas are given, and are included chiefly for use in preliminary estimates when the only data are the centre heights and the angles of the transverse surface slopes. With these, the heights of equivalent level sections are readily found by Mr. Trautwine's well-known and very ingenious diagrams, than which for the purpose intended probably no better means can be devised. When these heights have been ascertained, the use of the special Correction Tables in connection with those of level cuttings will reduce to a minimum the labor of computing the prismoidal contents. If further tables of level cuttings are considered necessary, the reader is referred to Mr. Trautwine's "Excavation and Embankment," or to the example given at the end of this work, by careful attention to which any required table may be written out with entire accuracy in a few hours. Special corrections for any side slopes may be obtained by Rule 12.

Not an inconsiderable advantage of the present method is that, by giving accurate corrections for the familiar approximations in general use, the calculator has the element of error constantly before him, and must speedily learn by practice, if not by theory, the cases in which such corrections become important. But while enough is given, both by rule and example, in Part II. to guide the least theoretical in the use of the tables, in Part I. a strictly mathematical investigation of principles and derivation of formulæ is submitted to the careful reader.

The article on Correction of Contents for Curvature was suggested by that on the same subject in "Henck's Field-Book," but, by the formulæ and table of factors given, in ordinary cases the corrections are much more readily obtained in practice.

All of the tables in this work have been calculated by the writer, and, as the system used was that of continued additions with special tests at intervals, it is believed that they will be found absolutely correct within the purposed limits, whether the last figure of any amount given be intended to express the nearest whole number or the nearest decimal.

NOTATION AND SIGNS USED.

A and A' = end areas of earthwork.
M = middle area.
a and a' = areas of triangle between road-bed and intersection of side slopes produced.
b and b' = road-bed widths.
c and c' = centre heights of profile.
h and h' = heights of equivalent level sections.
s and s' = ratios of opposite side slopes to 1.
d and d' = side distances.
h_1 and h_2 = side heights.
N, N', n and n' = correction numbers.
C = contents for 100 feet.
Q = correction for curvature.
$\gtrless$ = "greater or less than."
$\sim$ = "the difference between."
"Grade triangle" = triangle between the base and the intersection of the side slopes produced.

EARTHWORK MENSURATION.

PART I.

AREAS.—GROUND SLOPING TRANSVERSELY.   THOROUGH-CUT.

Fig. 1.

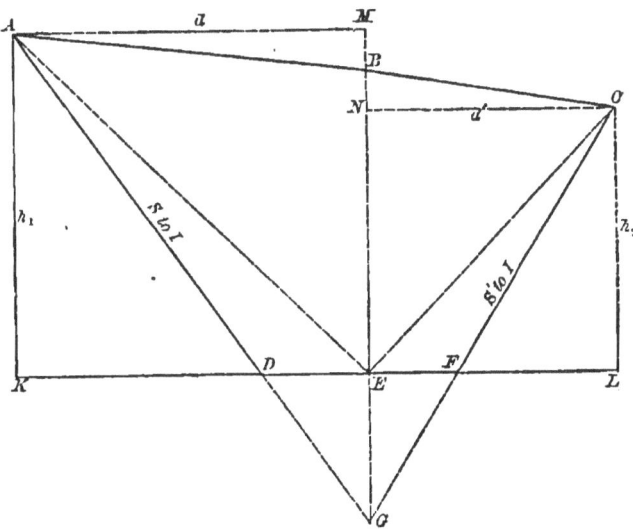

Let area ABCFD $= A$, area DFG $= a$, centre height BE $= c$, side heights AK and CL $= h_1$ and h_2, side distances AM and NC $= d$ and d', base DF $= b$, and ratios of side slopes to $1 = s$ and s'.

CASE 1.—Side slopes the same. $s' = s$. Produce the side slopes until they meet in G.

$$EG \times s = \frac{b}{2}, \text{ hence } EG = \frac{b}{2s}$$

$$\text{and area } a = \frac{b \times \frac{b}{2s}}{2} = \frac{b^2}{4s}$$

But $BG = c + \frac{b}{2s}$, hence

$$\text{area } ACG = A + a = \left(c + \frac{b}{2s}\right)\left(\frac{d+d'}{2}\right)$$

$$\text{and } A = \frac{\left(c + \frac{b}{2s}\right)(d+d')}{2} - \frac{b^2}{4s} \dots\dots\dots\dots(1)$$

Example.—Given $s' = s = \frac{3}{4}$; $b = 18$ ft.; $d = 30.9$; $d' = 21.6$; $c = 22.0$.

$$\frac{b}{2s}(\text{tab. 1}) = 12, \text{ and } a (\text{tab. 2}) = 108.$$

$$A + a = \frac{(22.0+12.0)(30.9+21.6)}{2} = 892.5$$

and $A = 892.5 - 108 = 784.5$.

CASE 2.—Opposite side slopes unequal. $s' \gtrless s$.

The areas of the triangles DAE, EAB, BCE, and ECF are respectively

$$\frac{\frac{b}{2} \times h_1}{2}, \frac{c \times d}{2}, \frac{c \times d'}{2}, \text{ and } \frac{\frac{b}{2} \times h_2}{2}$$

and, $A = \dfrac{\frac{b}{2}(h_1 + h_2) + c(d + d')}{2}$ ....(2)

Example.—$s = \frac{1}{4}$; $s' = 1$; $b = 16$; $c = 12.6$; $d \& d' = 10.1 \& 29.8$; $h_1 \& h_2 = 8.4 \& 21.8$.

$$A = \frac{8(8.4+21.8) + 12.6(10.1+29.8)}{2} = 370.6.$$

CASE 3.—DE greater or less than EF.

Let $DE = \dfrac{b}{2}$, and $EF = \dfrac{b'}{2}$

The triangles DAE, EAB and BCE have the same expressions for their areas as in case 2, and area $ECF = \dfrac{\frac{b'}{2} \times h_2}{2}$

hence, $A = \dfrac{\frac{bh_1}{2} + \frac{b'h_2}{2} + c(d+d')}{2}$ ....(3)

Example.—Double width track. $s = \frac{1}{2}$; $s' = \frac{3}{4}$; $\frac{b}{2} = 9$; $\frac{b'}{2} = 21$

$c = 32.8$; $h_1 \& h_2 = 24.4 \& 40.4$; $d \& d' = 21.2 \& 51.3$

$$A = \frac{9.0 \times 24.4 + 21.0 \times 40.4 + 32.8(21.2 + 51.3)}{2} = 1723$$

Formula (1) applies only to case 1; formula (2) to cases 1 and 2; and formula (3) is general for all cases where the whole road-bed width is either in cutting or embankment, and the surface slopes are sensibly regular between the centre and side stakes.

AREAS.—SIDE HILL CUTTING.

Let $q =$ the horizontal distance from centre line to grade point opposite, and $A =$ the area of excavation.

CASE 1.—Both centre and side height in excavation.

The areas of triangles DAE and EAB are as before, and that of the triangle running out to grade $= \frac{cq}{2}$

hence, $\quad A = \dfrac{\dfrac{bh_1}{2} + c(d+q)}{2}$ ....(4)

Example.—$s = 1$, $b = 20$, $c = 4.3$, $h_1 = 10.6$, $d = 20.6$, and $q = 6.2$.
$$A = \frac{10 \times 10.6 + 4.3\,(20.6 + 6.2)}{2} = 110.6$$

CASE 2.—Centre height in embankment.
$$A = \frac{\left(\dfrac{b}{2} - q\right)h_1}{2} \quad \ldots\ldots\ldots\ldots\ldots(5)$$

Example.—$b = 18$, $h = 10$, $q = 5$. $A = \dfrac{(9-5)\,10}{2} = 20$

AREAS.—GROUND LEVEL TRANSVERSELY.
Fig. 2.

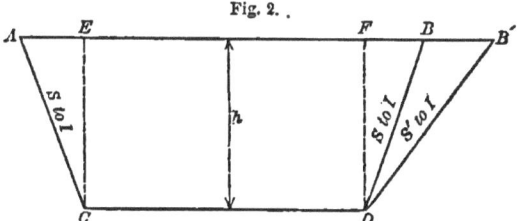

CASE 1.—Side slopes the same, or $s' = s$.

$AE = FB = hs$, and $EF = CD = b$

Area $ABCD = \left(\dfrac{AB + CD}{2}\right)h = \left(\dfrac{hs + b + hs + b}{2}\right)h$

or $A = (b + hs)h$ ....(6)

Example.— $s' = s = \dfrac{1}{2}$; $b = 16$; $h = 20$
$$A = \left(16 + 20 \times \tfrac{1}{2}\right)20 = 26 \times 20 = 520.$$

When the field notes are given, this example can, of course, be worked by any one of formulæ (1), (2), or (3).

CASE 2.—Opposite side slopes unequal, or $s' \gtrless s$.

$AE = hs$; $FB' = hs'$; and $EF = CD$.

area $AB'CD = \left(\dfrac{AB' + CD}{2}\right)h = \left(\dfrac{hs + b + hs' + b}{2}\right)h$

or $A = \left(b + h\left(\dfrac{s + s'}{2}\right)\right)h$ ....(7)

Example.—$s = \frac{1}{2}$; $s' = 1$; $b = 16$; $h = 20$.

$$A = \left(16 + 20 \times \frac{3}{4}\right)20 = 31 \times 20 = 620.$$

AREAS.—GROUND BROKEN TRANSVERSELY.
Fig. 3.

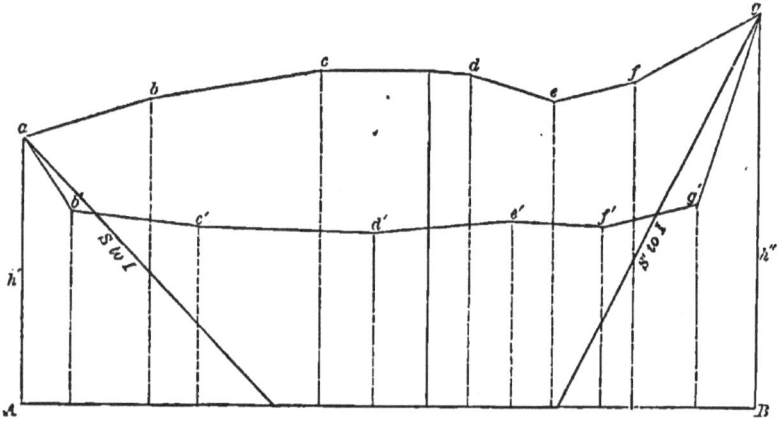

To calculate the area $abcdefg\ b'c'd'e'f'g$.

The elevations and horizontal distances apart of the points a, b, c, d, e, f, g, must be determined in the usual manner before the surface is disturbed, and of b', c', d', e', f', g', after the excavation is made.

Calculate the area $A\ a\ b\ c\ d\ e\ f\ g\ B$ between the surface line and the assumed datum plane AB; also

The area $A\ a\ b'\ c'\ d'\ e'\ f'\ g'\ g B$ between the bottom of the pit as excavated and the same datum plane AB.

The difference between the results so obtained, gives the area required.

When the cross sections of the line have the surface broken transversely, if the slope stakes are supposed to be at a and g (fig. 3), and AB is the plane of the road-bed, calculate

1st: the area $A\ a\ b\ c\ d\ e\ f\ g\ B$

2d: the triangles of excess $= \dfrac{h_1^2 s + h_2^2 s'}{2}$

The difference between the above two results will give the area of earthwork required.

For side hill work the process is similar, except that only one triangle of excess $= \dfrac{h_1^2 s}{2}$, is to be deducted.

This of course applies to embankment as well as excavation.

None of the preceding cases require that the cross section shall be drawn before calculating its area.

CONTENTS.—FRUSTUM FORMULA.

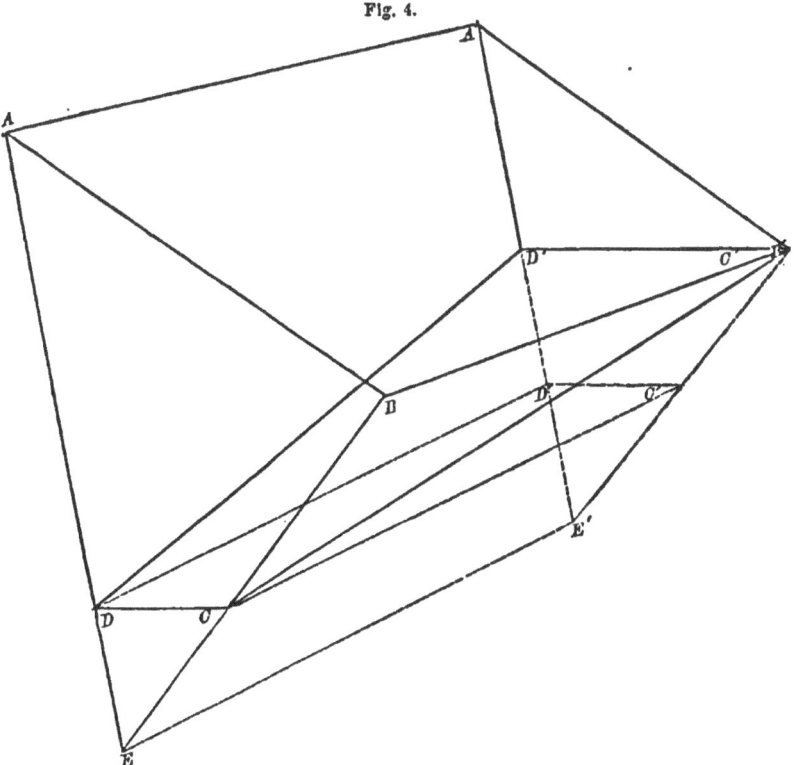

Fig. 4.

If ABCD and A'B'C'D' be two consecutive cross sections with like surface lines and side slopes but unequal bottom widths, by producing the side slopes until they meet at E and E', the whole figures ABE and A'B'E' are similar as well as the triangles CDE and C'D'E'. But the solid ABCDA'B'C'D' being the difference between the frustums ABEA'B'E' and CDEC'D'E' its cubic contents are

$$\left(ABE + A'B'E' + \sqrt{ABE \times A'B'E'}\right)\tfrac{l}{3}$$
$$- \left(CDE + C'D'E' + \sqrt{CDE \times C'D'E'}\right)\tfrac{l}{3}$$

in which l represents the distance between the cross sections.

If areas ABCD, A'B'C'D', CDE and C'D'E' be represented by A, A', a and a' respectively, then taking l as 100 feet, and representing the contents in cubic yards by C, we have:

$$C = \frac{(A+a)+(A'+a')+\sqrt{(A+a)(A'+a')}-(a+a'+\sqrt{aa'})}{3} \times \frac{100}{27}. \quad (8)$$

If CD = C'D' then $a' = a$, and the formula becomes:

$$C = \left(\frac{(A+a)+(A'+a)+\sqrt{(A+a)(A'+a)}}{3} - a\right)\frac{100}{27} \quad \ldots\ldots\ldots (9)$$

When CD = C'D' = 0, a vanishes, and

$$C = \left(\frac{A + A' + \sqrt{AA'}}{3}\right)\frac{100}{27} \quad \ldots\ldots\ldots\ldots\ldots (10)$$

which is the formula for the frustum of a pyramid.

By formulæ (8), (9), and (10) the whole of the formulæ for cubic contents hereafter given may be conveniently tested.

As the solid resulting from connecting the homologous sides of two similar and parallel sections of unequal areas is the frustum of a pyramid, formula (10) is applicable to any plane solid with such end sections.

CONTENTS.—PRISMOIDAL FORMULA.

Fig. 5.

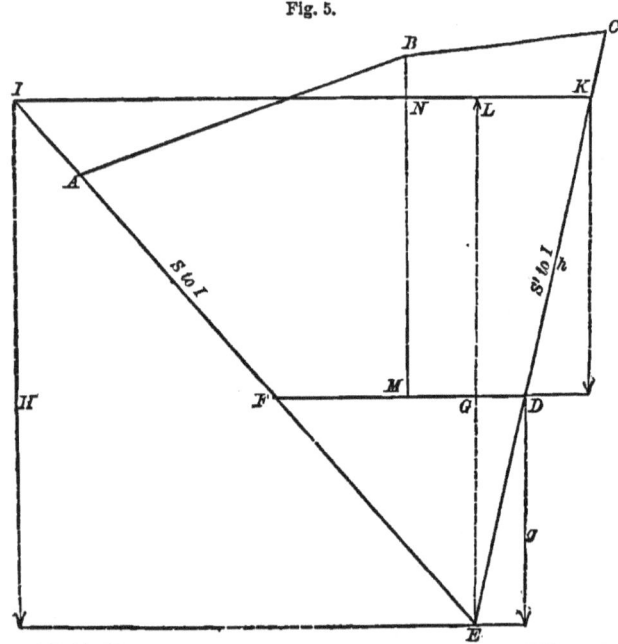

Let ABCDF be a given cross section, with a base FD = b, and s

and s' the ratios of its side slopes to 1; also let IKDF be an equivalent cross section with level surface, height $MN = h$, and with same base and side slopes. Produce the side slopes to their intersection at E, and from E let fall the perpendicular EL on IK, intersecting the base in G. Let area $ABCDF = IKDF = A$, and $FDE = a$.

In the triangle FDE, $FG = EG \times s$, and $GD = EG \times s'$, or $FD = EG(s+s')$, whence $EG = \dfrac{FD}{s+s'} = \dfrac{b}{s+s'}$ and area $FDE = \dfrac{FD \times EG}{2} = \dfrac{b}{2} \times \dfrac{b}{s+s'} = \dfrac{b^2}{2(s+s')} = a.$

Similarly in triangle IKE, $EL = h + \dfrac{b}{s+s'}$

$IK = \left(h + \dfrac{b}{s+s'}\right)(s+s')$, and area $IKE = \left(h + \dfrac{b}{s+s'}\right)^2 \left(\dfrac{s+s'}{2}\right) = A + a$;

consequently,

$$A = \overline{EL}^2\left(\dfrac{s+s'}{2}\right) - a = \left(h + \dfrac{b}{s+s'}\right)^2\left(\dfrac{s+s'}{2}\right) - \dfrac{b^2}{2(s+s')} \quad \ldots \ldots \ldots \ldots (11)$$

from which,

$$EL = h + \dfrac{b}{s+s'} = \sqrt{\left(A + \dfrac{b^2}{2(s+s')}\right)\dfrac{2}{s+s'}} = \sqrt{(A+a)\dfrac{2}{s+s'}}$$

For convenience of calculation, let $GE = \dfrac{b}{s+s'}$, be represented by g, and EL by H; then as $\dfrac{b^2}{2(s+s')} = \left(\dfrac{b}{s+s'}\right)^2 \dfrac{s+s'}{2} = g^2\left(\dfrac{s+s'}{2}\right)$

we have, by substitution in (11),

$$A = (H^2 - g^2)\dfrac{s+s'}{2}$$

For a second section with corresponding parts b', H$'$, s and s', and areas A$'$ and a'

$$A' = (H'^2 - g'^2)\dfrac{s+s'}{2}$$

and for the area M of a cross section midway between A and A$'$,

$$M = \left(\left(\dfrac{H+H'}{2}\right)^2 - \left(\dfrac{g+g'}{2}\right)^2\right)\dfrac{s+s'}{2} \ldots \ldots \ldots \ldots (12)$$

The prismoidal formula for the contents C between two end areas A and A$'$ at a distance apart $= l$, with an area M midway between them is:

$$C = \left(\dfrac{A + A' + 4M}{6}\right)l \ldots \ldots \ldots \ldots \ldots \ldots (13)$$

But $\dfrac{A+A'}{6} = \dfrac{A+A'}{2} - \dfrac{A+A'}{3}$

and by substitution in (13)

$$C = \left(\dfrac{A+A'}{2} - \dfrac{A+A'-2M}{3}\right)l \quad \ldots\ldots\ldots\ldots (14)$$

also $\dfrac{4M}{6} = M - \dfrac{2M}{6}$; and substituting this in (13)

$$C = \left(M + \dfrac{A+A'-2M}{6}\right)l \quad \ldots\ldots\ldots\ldots (15)$$

The two last expressions for the value of C show that the calculation of contents by averaging the end areas requires a *minus* correction; and by the middle area (or, what is equivalent, taking the amount corresponding to the average of the end heights from a special table) a *plus* correction of exactly half as much. The actual *minus* correction will now be found. By substituting the values of A, A' and M in the second term of (14) we have :

$$C = \left(\dfrac{A+A'}{2} - \dfrac{(H^2-g^2)\frac{s+s'}{2} + (H'^2-g'^2)\frac{s+s'}{2} - 2\left(\left(\frac{H+H'}{2}\right)^2 - \left(\frac{g+g'}{2}\right)^2\right)\frac{s+s'}{2}}{3}\right)l$$

and reducing*

$$C = \left(\dfrac{A+A'}{2} - \left(\dfrac{(H-H')^2 - (g-g')^2}{6}\right)\dfrac{s+s'}{2}\right)l \quad \ldots\ldots (16)$$

But $H = \sqrt{\left(A + \dfrac{b^2}{2(s+s')}\right)\dfrac{2}{s+s'}}$; $H' = \sqrt{\left(A' + \dfrac{b'^2}{2(s+s')}\right)\dfrac{2}{s+s'}}$;

$g = \dfrac{b}{s+s'}$; and $g' = \dfrac{b'}{s+s'}$, and by substitution in (16)

$$C = \left\{\dfrac{A+A'}{2} - \left(\dfrac{\left(\sqrt{\left(A + \frac{b^2}{2(s+s')}\right)\frac{2}{s+s'}} - \sqrt{\left(A' + \frac{b'^2}{2(s+s')}\right)\frac{2}{s+s'}}\right)^2 - \left(\frac{b-b'}{s+s'}\right)^2}{6}\right)\dfrac{s+s'}{2}\right\}l$$

---

* Neglecting the common factors $\dfrac{s+s'}{2}$ and l, and the denominator, the second term becomes,

$(H^2-g^2)+(H'^2-g'^2)-2\left(\dfrac{(H+H')^2}{4} - \dfrac{(g+g')^2}{4}\right) = H^2-g^2+H'^2-g'^2$

$\qquad\qquad\qquad\qquad\qquad -\dfrac{H^2+2HH'+H'^2}{2} + \dfrac{g^2+2gg'+g'^2}{2}$

$= \dfrac{2H^2-2g^2-2H'^2-2g'^2-H^2-2HH'-H'^2+g^2+2gg'+g'^2}{2}$

$= \dfrac{H^2-2HH'+H'^2-g^2+2gg'-g'^2}{2} = \dfrac{(H-H')^2-(g-g')^2}{2}$

and restoring the factors $\dfrac{s+s'}{2}$ and l, and the denominator, we obtain formula (16).

Reducing :*

$$C = \left\{ \frac{A+A'}{2} - \underbrace{\left(\sqrt{A+\frac{b^2}{2(s+s')}} - \sqrt{A'+\frac{b'^2}{2(s+s')}}\right)^2 - \left(\frac{b-b'}{s+s'}\right)^2 \frac{s+s'}{2}}_{6} \right\} l$$

making $l = 100$, dividing by 27, observing that $(x-y)^2 = (y-x)^2 = (y\sim x)^2$, and that $\frac{b^2}{2(s+s')} = a$, we obtain :

$$C = \left(\frac{A+A'}{2} - \frac{(\sqrt{A+a} \sim \sqrt{A'+a'})^2}{6} + \frac{(b\sim b')^2}{\frac{2(s+s')}{6}}\right)\frac{100}{27} \quad \ldots\ldots\ldots (17)$$

This is the general formula when the opposite side slopes and end road-bed widths are both different.

When the road-bed widths are the same, or $b \sim b' = 0$, the last term vanishes, and the formula becomes:

$$C = \left(\frac{A+A'}{2} - \frac{(\sqrt{A+a} \sim \sqrt{A'+a})^2}{6}\right)\frac{100}{27} \quad \ldots\ldots\ldots\ldots (18)$$

This is the general formula for all slopes and bases where the base is constant between the two end sections.

When $b = b' = o$, $a = o$, and

$$C = \left(\frac{A+A'}{2} - \frac{(\sqrt{A} \sim \sqrt{A'})^2}{6}\right)\frac{100}{27} \quad \ldots\ldots\ldots\ldots\ldots (19)$$

This is the general formula for the frustum of a pyramid,† such as may be the solid between two sections of side hill excavation.

The correction in terms of equivalent level heights h and h' may be found directly from (16) as follows :

When $b' = b$, the expression $(g-g')^2$ vanishes and (16) becomes :

---

* In squaring the binomial of radicals the factor $\sqrt{\frac{2}{s+s'}}$ becomes $\left(\sqrt{\frac{2}{s+s'}}\right)^2$ in the first term, $\sqrt{\frac{2}{s+s'}}\sqrt{\frac{2}{s+s'}}$ in the second, and $\left(\sqrt{\frac{2}{s+s'}}\right)^2$ in the third, or in each $\frac{2}{s+s'}$, thus cancelling the factor $\frac{s+s'}{2}$, except in the last term of the numerator.

† Formula (10) before given for the frustum of a pyramid may be transformed into formula (19); for $\frac{A+A+\sqrt{AA'}}{3} - \frac{2A+2A'+2\sqrt{AA'}}{6} =$
$\frac{3A+3A'-A-A'+2\sqrt{AA'}}{6} - \frac{3(A+A')}{6} - \frac{A-2\sqrt{AA'}+A'}{6} - \frac{A+A'}{2} -$
$\frac{(\sqrt{A} \sim \sqrt{A'})^2}{6}$. When $A'=0$ in formula (19) it becomes $C = \left(\frac{A}{2} - \frac{(\sqrt{A})^2}{6}\right)\frac{100}{27}$
$= \left(\frac{A}{2} - \frac{A}{6}\right)\frac{100}{27} = \frac{A}{3} \times \frac{100}{27}$, which is the formula for the solidity of a pyramid, as it should be.

$$C = \left(\frac{A+A'}{2} - \frac{(H-H')^2}{6}\left(\frac{s+s'}{2}\right)\right)l$$

but $(H-H')^2 = \left(\left(h+\frac{b}{s+s'}\right)-\left(h'+\frac{b}{s+s'}\right)\right)^2 = (h-h')^2 = (h\sim h')^2$

and substituting, making $l = 100$, and dividing by 27,

$$C = \left(\frac{A+A'}{2} - \frac{(h\sim h')^2}{6}\left(\frac{s+s'}{2}\right)\right)\frac{100}{27} \quad\ldots\ldots\ldots\ldots\ldots(20)$$

As the plus correction for calculation by middle area was found to be one half of the minus correction for averaging end areas, by making the requisite changes in (20):

$$C = \left(M + \frac{(h\sim h')^2}{12}\left(\frac{s+s'}{2}\right)\right)\frac{100}{27}$$

but when $b'=b$, from formula (12), we obtain*

$$M = b\left(\frac{h+h'}{2}\right) + \left(\frac{h+h'}{2}\right)^2\frac{s+s'}{2}$$

and by substitution:

$$C = \left\{ b\left(\frac{h+h'}{2}\right) + \left(\left(\frac{h+h'}{2}\right)^2 + \left(\frac{h\sim h'}{12}\right)^2\right)\frac{s+s'}{2}\right\}\frac{100}{27}\ldots(21)$$

This formula is for use when the equivalent level heights have been obtained.

APPLICATION OF THE PRISMOIDAL FORMULA.

The prismoidal formula in its ordinary form is applicable to a variety of solids, regular and irregular, but requires that the actual middle section shall be previously determined and its area known.

In a modified form it can be applied practically by means of tables; such applications, however, always involving a value of the

---

* By substituting the values of H, H', g and g' in formula (12) it becomes:

$$M = \left(\frac{\left(h+\frac{b}{s+s'}\right)+\left(h'+\frac{b'}{s+s'}\right)}{2}\right)^2 - \left(\frac{\frac{b}{s+s}+\frac{b'}{s+s'}}{2}\right)^2$$

making $b'=b$, and squaring:

$$M = \frac{\left(h+\frac{b}{s+s'}\right)^2 + 2\left(h+\frac{b}{s+s'}\right)\left(h'+\frac{b}{s+s'}\right) + \left(h'+\frac{b}{s+s'}\right)^2 - 4\left(\frac{b}{s+s'}\right)^2}{4}$$

$$= \frac{h^2 + \frac{2bh}{s+s'} + \left(\frac{b}{s+s'}\right)^2 + 2hh' + \frac{2bh'}{s+s'} + \frac{2bh}{s+s'} + 2\left(\frac{b}{s+s'}\right)^2 + h'^2 + \frac{2bh'}{s+s'} + \left(\frac{b}{s+s'}\right)^2 - 4\left(\frac{b}{s+s'}\right)^2}{4}\left(\frac{s+s'}{2}\right)$$

$$= \frac{2bh\left(\frac{2}{s+s'}\right) + 2bh'\left(\frac{2}{s+s'}\right) + h^2 + 2hh' + h'^2}{4}\left(\frac{s+s'}{2}\right) = b\left(\frac{h+h'}{2}\right) + \left(\frac{h+h'}{2}\right)^2\frac{s+s'}{2}.$$

This also results directly from formula (7) by taking the area of a second section for a height of h', and averaging like parts for M.

middle area which can be deduced directly from the end *areas* without necessitating a previous knowledge of the parts of either the middle or the end *sections*.

But in all of its modifications, as well as in its ordinary form, the prismoidal formula invariably involves the area of the actual middle section of the solid to which it is applied, and, as in "Roots and Squares" and "Equivalent level heights," both methods involve a value of the area of this middle section (carried to intersection of side slopes when in thorough-cut) which can be proved identical with that of the frustum of a pyramid, the theoretical application of these methods is limited to solids with end sections sensibly similar, or which can be rendered so by being carried to the intersection of the side slopes.

As the above has been ignored by other writers on this subject, its mathematical proof will be given.

The contents of a frustum may be expressed either by the prismoidal or the frustum formula, therefore in the case of a frustum :

$$\frac{A+A'+4M}{6} \times l = \frac{A+A'+\sqrt{AA'}}{3} \times l$$

whence $A+A'+4M = 2A+2A'+2\sqrt{AA'}$, and $M = \dfrac{A+A'+2\sqrt{AA'}}{4}$

$$= \left(\frac{\sqrt{A'}+\sqrt{A'}}{2}\right)^2$$

The formula of Roots and Squares where A and A′ represent the end sections* is (Formula 19) :

$$C = \left(\frac{A+A'}{2} - \frac{(\sqrt{A}-\sqrt{A'})^2}{6}\right)\frac{100}{27}$$

and the prismoidal formula for the same solid is :

$$C = \left(\frac{A+A'+4M}{6}\right)\frac{100}{27}$$

hence $\dfrac{A+A'+4M}{6} = \dfrac{A+A'}{2} - \dfrac{(\sqrt{A}-\sqrt{A'})^2}{6}$

clearing fractions, $A+A'+4M = 3A+3A'-(\sqrt{A}-\sqrt{A'})^2$

and $M = \dfrac{2A+2A'-A+2\sqrt{AA'}-A'}{4} = \left(\dfrac{\sqrt{A}+\sqrt{A'}}{2}\right)^2$

In two end sections with surface level transversely and side slopes constant, if H and H′ represent the heights from intersection of side slopes to surface and s the ratio of the side slopes to 1, the areas of

---

* In this article, whether the end sections are carried to intersection of side slopes or not, their areas are expressed by A and A′.

the end sections to intersection are $H^2s = A$, and $H'^2s = A'$, and for the area of the middle section, by averaging like parts:

$$M = \left(\frac{H+H'}{2}\right)^2 s = \left(\frac{H\sqrt{s}+H'\sqrt{s}}{2}\right)^2 = \left(\frac{\sqrt{H^2s}+\sqrt{H'^2s}}{2}\right)^2$$

$$= \left(\frac{\sqrt{A}+\sqrt{A'}}{2}\right)^2$$

which is the same value of M as that before obtained. Substituting this in the prismoidal formula:

$$C = \frac{A+A'+4\left(\frac{\sqrt{A}+\sqrt{A'}}{2}\right)^2}{6} \times \frac{100}{27}, \text{ and reducing,}$$

$$C = \frac{A+A'+A+2\sqrt{AA'}+A'}{6} \times \frac{100}{27} = \frac{A+A'+\sqrt{AA'}}{3} \times \frac{100}{27}$$

which is the formula for the frustum of a pyramid, and shows that this value of M introduced into the prismoidal formula limits its application to such solids only as are frustums of pyramids. This will be illustrated further from Example 5, page 36, in which when carried to the intersection of the side slopes produced, the end sections are similar.

Thus carried to intersection, the end areas and the actual middle area are respectively 349, 2951, and 1333, as given page 36.

By Roots and Squares

$$M = \left(\frac{\sqrt{349}+\sqrt{2951}}{2}\right)^2 = 1332$$

By equivalent level heights

$$H = \sqrt{\frac{A}{s}} = \sqrt{349 \times \tfrac{2}{3}} = 15.25$$

$$H' = \sqrt{\frac{A'}{s}} = \sqrt{2951 \times \tfrac{2}{3}} = 44.35$$

$$M = \left(\frac{H+H'}{2}\right)^2 s = \left(\frac{15.25+44.35}{2}\right)^2 \times \tfrac{3}{2} = 1332$$

By substituting this value of M in the prismoidal formula:

$$C = \frac{349+2951+4\times 1332}{6} \times \frac{100}{27} = 1438 \text{ tab. } 4 = 5326 \text{ cyds.}$$

For calculation by equivalent level heights as table 15 has a base of 14 feet, and the above heights are taken to intersection of side slopes, $\left(\frac{H+H'}{2}\right) \times 14 \times \frac{100}{27}$ must be deducted from contents taken from tables.

By Rule 4,
$$\frac{15.25+44.35}{2} = 29.8 \text{ table } 15..6,479$$
$$15.25 \sim 44.35 = 29.1 \text{ table } 17..+392$$
$$\overline{6,871}$$
$$\text{Deduct } 29.8 \times 14 \times \frac{100}{27} = 417.2 \text{ table } 4... -1,545$$
$$\overline{5,326 \text{ cyds.}}$$

By mean proportional or frustum formula:
$$C = \frac{349+2951+\sqrt{349 \times 2951}}{3} \times \frac{100}{27} = 1438.3 \text{ table } 4...5,327 \text{ cyds.}$$

By deducting the grade prism $32.7 \times \frac{100}{27} = 121$ cyds., practically the same result as that given on page 36 is obtained.

Another case in which the area of the actual middle section can be deduced from the end areas directly, is when each of the latter can be expressed by two surface dimensions, one of which is the same for both end sections, as in solids whose end sections are parallelograms or triangles with the same base and different heights, or *vice versa*. Thus if $bh = A$ and $bh' = A'$ represent the end areas of a solid of which the end sections are triangles with the same base and different heights, as may be the case in side hill cutting where the transverse surface slope increases regularly between the end sections, by averaging like parts the middle area is

$$M = b\left(\frac{h+h'}{2}\right) = \frac{bh+bh'}{2} = \frac{A+A'}{2}$$

And as the prismoidal formula is applicable here, by substituting this value of M:

$$C = \frac{A+A'+\left(\frac{A+A'}{2}\right)4}{6} \times \frac{100}{27} = \frac{A+A'}{2} \times \frac{100}{27}$$

which is the average area formula, in this case giving the prismoidal contents. As an example, suppose the triangular end sections of the solid to have a base of 20 feet and heights of 10 and 40 feet respectively. Then $A = 10 \times 10 = 100$; $A' = 10 \times 40 = 400$; and $M = 10 \times \frac{10+40}{2} = 250 = \frac{A+A'}{2}$.

By the prismoidal formula:
$$C = \frac{100+400+4 \times 250}{6} \times \frac{100}{27} = 250 \text{ table } 4...926 \text{ cyds.}$$

Calculated by Roots and Squares $M = \left(\dfrac{\sqrt{100}+\sqrt{400}}{2}\right)^2 = 225$,

and this substituted in the prismoidal formula gives

$$C = \dfrac{100+400+4\times 225}{6} \times \dfrac{100}{27} = 233.3 \text{ table } 4 = 864 \text{ cyds.}$$

Here the average area formula gives the prismoidal contents, and the prismoidal formula applied by its modification of Roots and Squares gives a very rough approximation. The same inaccuracy is of course involved in the method by equivalent level heights, whatever may be the shape of the equivalent and similar end sections of which the level heights are obtained. For instance, if the side hill work is excavated at rock slope, the level heights, if carried to vertex, may be taken for sections with any other side slopes, as 1 to 1, or 1¼ to 1.

At 1 to 1 carried to vertex $H = \sqrt{\dfrac{100}{1}} = 10$; $H' = \sqrt{\dfrac{400}{1}} =$ 20, and to calculate by table 12, with side slopes 1 × 1 and base 18 feet:

$$\dfrac{10+20}{2} = 15 \text{ table } 12\ldots\ldots\ldots\ldots 1833$$

$$10 \sim 20 = 10 \text{ table } 14\ldots\ldots\ldots\ldots +31$$

Deduct $15\times 18 \times \dfrac{100}{27} = 270 \text{ table } 4\ldots\ldots\ldots\ldots -1000$

864 cyds.

at 1¼ to 1 carried to vertex $H = \sqrt{100\times\frac{4}{5}} = 8.16$; $H' = \sqrt{400\times\frac{2}{3}}$ = 16.33, and to calculate by table 15, with side slopes 1¼ to 1, and base 14 feet.

$$\dfrac{8.16+16.33}{2} = 12.245 \text{ table } 15\ldots\ldots 1468$$

$$8.16 \sim 16.33 = 8.17 \text{ table } 17\ldots\ldots +31$$

Deduct $12.245 \times 14 \times \dfrac{100}{27} = 171.4 \text{ table } 4\ldots\ldots -635$

864 cyds.

The two last examples show the same error of 62 cyds. obtained by Equivalent level heights, as before by Roots and Squares.

By mean proportionals or frustum formula:

$$\dfrac{100+400+\sqrt{100\times 400}}{3} \times \dfrac{100}{27} = 233.3 \text{ table } 4\ldots\ldots 864 \text{ cyds.}$$

By Rule 2,

$$\frac{100+400}{2} = 250 \text{ table 4}\ldots\ldots\ldots\ldots926$$
$$10\sim20 = 18 \text{ table 5}\ldots\ldots\ldots\ldots\ 62$$
$$\overline{864 \text{ cyds.}}$$

If the above sections were similar, as for instance with dimensions 10 × 10 and 20 × 20, the first method by average areas would give too much by 62 cyds, whilst by the others the true prismoidal contents would be obtained.

If both the heights and bases are different and the sections are not similar, the middle area will be less than $\frac{A+A'}{2}$ and greater than $\left(\frac{\sqrt{A}+\sqrt{A'}}{2}\right)^2$, and cannot be obtained directly from the end areas. In such cases, the exact contents can be determined by the prismoidal formula only by first obtaining the dimensions of the actual middle section and calculating its area.

Practically in railroad earthwork it is only when the transverse surface lines of the end sections are very dissimilar and the areas differ greatly in size that the resulting errors become important, and as at such points the cross sections are usually taken nearer together, it is very rarely the case that the methods of Roots and Squares and Equivalent level heights fail of practical correctness. In cases of doubt, however, especially when the surface is warped between the end sections, it is safer and better to obtain the area of the actual middle section before calculating the contents.

CORRECTION OF CONTENTS FOR CURVATURE.

The following article was suggested by that given in Henck's "Field Book," page 110.

In excavation on curves, although the cross sections are actually staked out in the direction of the radii at the extremities of the chords, the calculation of contents is made as if these cross sections were perpendicular to the chords. In some cases, especially where the transverse surface slope is considerable, this is the occasion of a sensible error requiring a corresponding correction, the amount of which is determined as follows :

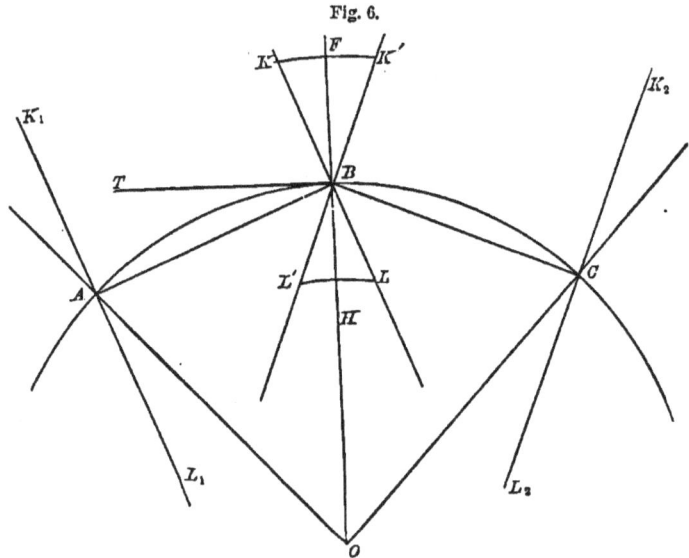

Fig. 6.

Suppose A, B, and C to be three consecutive 100 feet stations on a curve of radius OB; and BF and BH the side distances at station B.

The calculation of contents between A and B, and B and C made as if the cross sections at these points were on the lines K_1L_1, and KL, and $K'L'$ and K_2L_2, or perpendicular to the chords AB and BC, requires at each station a correction similar to that at B, which will now be considered. It is evident that the correction is the difference between the masses KBK' and L'BL, on opposite sides of the centre line, and between the two vertical planes KL and K'L'; these masses having for their cross sections respectively the half-breadths BF and BH. The angle KBK' being very small, the arcs KFK' and L'HL will be considered as straight lines; and, as the angle KBF = L'BH = $\frac{1}{2}$ KBK' = TBA = D, the deflection angle of the curve, the distance KF = BF × sin D; or, generally for small angles, any horizontal line as KK' or L'L measured perpendicularly to the radius OB, and terminated by the planes KL and K'L', is practically equal to BF or BH (the corresponding horizontal distance from the centre line) multiplied by 2 sin D. Consequently, the masses KBK' and L'BL being considered as truncated prisms with the areas of the half-breadths BF and BH as bases, their heights at any given points are equal to the horizontal distances of these points from the centre line, multiplied into twice the sine of the deflection angle.

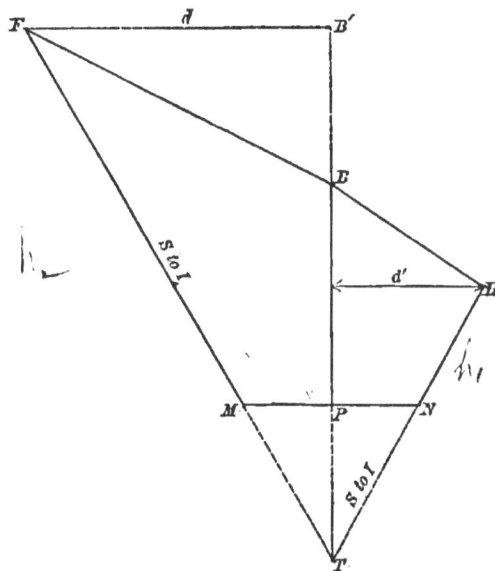

Fig. 7.

Conditions.—Single width road-bed and opposite side slopes the same. Transverse surface slopes regular.

Let FBHT represent the cross section at B (Fig. 6).
To simplify calculations, the equal prisms MPT and PTN are added.

The area FBT = $(BP+PT)\dfrac{FB'}{2} = \left(c+\dfrac{b}{2s}\right)\dfrac{d}{2}$, and the heights of the prism corresponding are $= d \times 2 \sin D$ at F, and $= 0$ at B and T. Its contents therefore $= \left(c+\dfrac{b}{2s}\right)\dfrac{d}{2}\times\left(\dfrac{d\times 2\sin D}{3}\right)$. Similarly the contents of prism HBT $= \left(c+\dfrac{b}{2s}\right)\dfrac{d'}{2}\times\left(\dfrac{d'\times 2\sin D}{3}\right)$ and the correction required, which is the difference of their volumes,

$$= \left(c+\dfrac{b}{2s}\right)\dfrac{d^2}{2}\times\dfrac{2\sin D}{3} \sim \left(c+\dfrac{b}{2s}\right)\dfrac{d'^2}{2}\times\dfrac{2\sin D}{3}$$

$$= \left(c+\dfrac{b}{2s}\right)\left(\dfrac{d^2\sim d'^2}{2}\right)\left(\dfrac{2\sin D}{3}\right)$$

and if Q represents the required correction in cubic yards,

$$Q = \left(c+\dfrac{b}{2s}\right)\left(\dfrac{d+d'}{2}\right)(d\sim d')\left(\dfrac{2\sin D}{3\times 27}\right) \dots \dots (22)$$

But, from formula (1), $\left(c+\dfrac{b}{2s}\right)\left(\dfrac{d+d'}{2}\right) = A+a$, the area carried to intersection of side slopes; also $\sin D = \dfrac{50}{R}$, and as $R = \dfrac{5730}{C°}$, in

which C° represents the degree of curve, $2 \sin D = 50 \times 2 \times \frac{C°}{5730}$
$= \frac{C°}{57.3}$

Therefore,
$$Q = (A+a) \, C° \times \frac{(d \sim d')}{57.3 \times 3 \times 27} \quad \ldots \ldots \ldots \ldots \ldots \ldots (23)$$

In side hill work, as shown by Mr. Henck, the general formula for the correction in cubic feet is $Q = \frac{wh}{2}(d+b-w)\frac{100}{3R}$, in which w represents the width of excavation at the road-bed. But as $\frac{wh}{2}$ = A, the area of earthwork, in this case the correction in cubic yards is

$$Q = A \times C° \times \frac{(d+b-w)}{57.3 \times 3 \times 27} \quad \ldots \ldots \ldots \ldots \ldots \ldots (24)$$

Values of the last factor in formulæ (23) and (24) are given in Table 18.

In excavation the correction for curvature as obtained by formulæ (23) and (24) is to be added when the curve is convex, and subtracted when it is concave toward the higher ground, and in embankment these conditions are reversed. It is supposed to be applied at the middle one of three cross sections at intervals of 100 feet, and all on the same curve.

If the distance to either of the cross sections next the one under consideration differs from 100 feet, the correction found as above is to be multiplied by the half sum of the two distances and divided by 100.

At points of curve or tangent one of these distances of course becomes nothing.

Whether the side slopes, or the widths from the centre line to the edge of the road-bed, are different or not, if the transverse surface lines are broken, the cross sections should be drawn to scale, the two half-breadths divided into triangles, and the horizontal distances from the centre line to the corners of each subdividing triangle measured on the drawing. The sum of the three distances for each triangle multiplied by its area and by $\frac{2 \sin D}{3}$ will give the contents in cubic feet of the prism corresponding. It is not material how the sides of the subdividing triangles are drawn, provided that the whole of each triangle is on the same side of the centre line. The difference of the masses whose cross sections are the half-breadths FB and BH (Fig. 6), and which lie on opposite sides of the centre line between the vertical planes KL and K′L′, the base plane and the planes of the side slopes, is in all cases the correction required.

With double-width track or opposite side slopes different, if the surface is regular from the centre to the slope stakes, from formula (3), the areas of the triangles of one half-breadth are $\frac{b}{4} \times h_1$ and $\frac{cd}{2}$, and of the other $\frac{b'}{4} \times h_2$ and $\frac{cd'}{2}$

The heights of the prisms corresponding to these areas are $\left(d+\frac{b}{2}+0\right)\frac{2}{3}\sin D$; $(d+0+0)\frac{2}{3}\sin D$; $\left(d'+\frac{b'}{2}+0\right)\frac{2}{3}\sin D$; and $(d'+0+0)\frac{2}{3}\sin D$, and their contents $\left(\frac{b}{4}\times h_1\right)\left(d+\frac{b}{2}\right)\frac{2}{3}\sin D$; $\left(\frac{cd^2}{2}\right)\frac{2}{3}\sin D$; $\left(\frac{b'}{4}\times h_2\right)\left(d'+\frac{b'}{2}\right)\frac{2}{3}\sin D$; and $\left(\frac{cd'^2}{2}\right)\frac{2}{3}\sin D$; but as $\frac{2}{3}\frac{\sin D}{27} = C° \times 0.000215$, the correction in cubic yards becomes

$$Q = \left\{\left(\frac{b}{4}\times h_1\right)\left(d+\frac{b}{2}\right) \sim \left(\frac{b'}{4}\times h_2\right)\left(d'+\frac{b'}{2}\right) + c\left(\frac{d+d'}{2}\right) \times (d \sim d')\right\} C° \times 0.000215 \ldots\ldots\ldots(25)$$

PART II.

PLAIN INSTRUCTIONS
FOR OBTAINING THE PRISMOIDAL CONTENTS OF EARTHWORK, WITH PRACTICAL RULES AND EXAMPLES SHOWING THE USES OF THE ACCOMPANYING TABLES IN SIMPLIFYING COMPUTATIONS BY THE FORMULÆ OF PART I.

The following Rules for computation of Cubic Contents are based on the condition that the transverse surface lines of the end sections shall be sensibly similar; but it will be observed that 1, 2, and 3 together cover all cases to which the method of "Roots and Squares," or of "Equivalent level heights," can be correctly applied, and that the practical limit of their application may be indefinitely extended by increasing the proximity of the cross sections in rough ground.

To find the prismoidal contents of thorough-cut or fill when road-bed width and side slopes are constant between end sections.

Given : areas, side slopes, and base (A and A', s and s', and b).

RULE 1.—(FORMULA 18).

Enter table 2 with the given road-bed width (b), and the half sum of the ratios of the side slopes $\left(\frac{s+s'}{2}\right)$, and take out the corresponding area $= a$. Add this to each of the given end areas and the square roots of the resulting quantities $\left(\sqrt{A+a} \text{ and } \sqrt{A'+a}\right)$ from table 3 are N and N', the correction numbers.

Enter table 4 with the average of the end areas $\left(\frac{A+A'}{2}\right)$, and table 5 with the difference of the correction numbers (N∼N'), and take out the corresponding quantities. The difference of the quantities taken from tables 4 and 5 is the contents in cubic yards for a length of 100 feet.

For a different length multiply by the length in feet and divide by 100.

Example.—Given $A = 974$; $A' = 87$; $s = \frac{1}{2}$; $s' = \frac{3}{4}$; $b = 20$.

From table 2 when $b = 20$ and $\frac{s+s'}{2} = \frac{4}{5}$, the area of the grade triangle $(a) = 160$

$\sqrt{A+a} = \sqrt{974 + 160} = 1134$ table 3 .... $33.7 = N$
$\sqrt{A'+a} = \sqrt{87 + 160} = 247$ table 3 .... $15.7 = N'$
$\frac{A+A'}{2} = \frac{974+87}{2} = 530.5$ table 4 1965
$N \sim N' = 33.7 \sim 15.7 = 18.0$ table 5 .... -200

Contents for 100 feet .... $\overline{1765}$ cyds.

For a different length as 80 feet, $1765 \times 0.8 = 1412$ cyds.

NOTE.—If the square roots of the areas to the intersection of the side slopes are obtained and recorded when the areas are calculated, as will ordinarily be found more convenient, the data are A and A' and N and N', and only the two last steps of Rule 1 are necessary.

To find the prismoidal contents of side hill work, pyramids, and any solid with similar end sections.

Given : end areas (A and A').

RULE 2 (FORMULA 19).

Take the square roots of the end areas $\left(\sqrt{A} \text{ and } \sqrt{A'}\right)$ from table $3 = n$ and n'.

Enter table 4 with the average of the end areas $\left(\frac{A+A'}{2}\right)$, and table 5 with the difference of the correction numbers $(n \sim n')$, and take out the corresponding quantities. The difference between the quantities taken from tables 4 and 5 is the contents in cubic yards for 100 feet.

For a different length multiply by the length in feet and divide by 100.

Example.—Given end areas $A = 41$ and $A' = 185$.

$\sqrt{A} = 41$ table $3 = 6.4 = n$; $\sqrt{A'} = 185$ table $3 = 13.6 = n'$.
$\frac{A+A'}{2} = \frac{41+185}{2} = 113$ table 4 .... 418.5
$n \sim n' = 6.4 \sim 13.6 = 7.2$ table 5 .... 32.0

Contents for 100 feet .... $\overline{386.5}$ cyds.

For a different length, as 25 feet, $\frac{386.5}{4} = 96.6$ cyds.

Example.—Pyramid. Given end areas $A = 104$ and $A' = 0$.
$\sqrt{A} = 104$ table $3 = 10.2 = n$; $\sqrt{A'} = 0 = n'$.

$$\frac{A+A'}{2} = \frac{104+0}{2} = 52 \text{ table } 4\ldots\ldots\ldots 192.6$$

$$n \sim n' = 10.2 \sim 0 = 10.2 \text{ table } 5 \ldots\ldots -64.2$$

Contents for 100 feet............128.4 cyds.

For a different length, as 60 feet, $128.4 \times 0.6 = 77$ cyds.

NOTE.—Examples under Rule 1 can be readily tested by Rule 2, the difference in the working being that the grade prism is first included and then deducted. For instance, in the example given under Rule 1, the end areas to intersection of side slopes are 1134 and 247, and the square roots corresponding 33.7 and 16.7—then :

$$\frac{1134+247}{2} = 695.5 \text{ table } 4\ldots\ldots\ldots 2558$$

$$33.7 \sim 15.7 = 18.0 \text{ table } 5 \ldots\ldots\ldots -200$$

Contents to intersection of side slopes...2358
Less grade prism 160 table 4..........−593

Contents of earthwork for 100 feet..1765 cyds.

To find the prismoidal contents of thorough-cut or fill when the end road-bed widths are different.

Given : end areas, side slopes, and end road-bed widths (A and A'; s and s' ; b and b').

RULE 3 (FORMULA 17).

Enter table 2 with $\frac{s+s'}{2}$ and b, b' and $b \sim b'$ respectively, and take out the corresponding areas a, a' and a''. From table 3 take out the square roots of the end areas to intersection $\sqrt{A+a} = N$, and $\sqrt{A'+a'} = N'$.

Enter table 4 with $\frac{A+A'}{2} + \frac{a''}{6}$, and table 5 with $N \sim N'$, and the difference between the corresponding quantities taken from tables 4 and 5 is the contents in cubic yards for 100 feet. For a different length multiply by the length in feet and divide by 100.

Example.—Given $b = 16$; $b' = 40$; $s = \frac{1}{4}$; $s' = \frac{3}{4}$; $A = 1565$; $A' = 253$.

Here $a = 128$; $a' = 800$; $a'' = 288$; $N = 41.1$ and $N' = 32.4$.

$$\frac{A+A'}{2} + \frac{a''}{6} = \frac{1565+253}{2} + \frac{288}{6} = 957 \text{ table } 4\ldots\ldots 3544.4$$

$$N \sim N' = 41.1 \sim 32.4 = 8.7 \text{ table } 5\ldots\ldots\ldots -46.7$$

Contents for 100 feet..................3497.7

For a different length, as 50 feet $\cdots\cdots \frac{3497.7}{2} = 1749$ cyds.

The example under Rule 3 is of a case where averaging the end areas gives less than the prismoidal contents. It may be tested by Formula 8, page 12, as also Rules 1 and 2 by Formulæ 9 and 10.

To find the prismoidal contents when the ground is level transversely, or where the heights of equivalent level sections have been obtained.

Given : level heights, base and half-sum of ratios of side slopes $\left(h \text{ and } h'\,;\ b \text{ and } \frac{s+s'}{2}\right)$.

RULE 4 (FORMULA 21).

Enter the table of level cuttings for the proper base and side slopes with the half-sum of the end heights $\left(\frac{h+h'}{2}\right)$, and the table of special plus corrections for the same side slopes with the difference of the end heights ($h \sim h'$), and take out the corresponding quantities. The sum of these quantities is the contents for 100 feet.

For a different length, multiply by the length in feet and divide by 100.

Example.—Given $b = 14$; $h = 8.6$; $h' = 36.8$; $\frac{s+s'}{2} = 1\frac{1}{4}$.

$\frac{h+h'}{2} = \frac{8.6+36.8}{2} = 22.7$ table 15.............. 4040

$h \sim h' = 8.6 \sim 36.8 = 28.2$ table 17............ +368

Contents for 100 feet................... 4408 cyds.

For a different length, as 85 feet, $4408 \times 0.85 = 3747$ cyds.

To find the Correction for Curvature in single width thorough-cut when the transverse surface slope is regular.

Given : area to intersection of side slopes, degree of curve, and difference of side distances (A+a, C°, and $d \sim d'$).

RULE 5 (FORMULA 23).

Enter table 18 with $d \sim d'$ and take out the corresponding factor : multiply this into the product of A+a by C°, and the result is Q the correction in cubic yards, to be applied at the middle one of three stations, all on the same curve and 100 feet apart. If the distance to either of the other two stations from the middle one differs from 100 feet, multiply by the half-sum of the two distances and divide by 100.

This correction is to be *added* or *subtracted* accordingly as the curve is *convex* or *concave* toward the higher ground.

Example.—Given $c = 28$; $h_1 = 40$; $h_2 = 16$; $d = 74$; $d' = 38$; $b = 28$; $R = 1400$; or $A+a = 2090$; $C° = 4°.09$; $d\sim d' = 36$.
$$36 \text{ table } 18 = 0.00776,$$
and $2090 \times 4.09 \times 0.00776 = 66.3$ cyds.

If the distances to the two adjacent stations are 50 and 40 feet respectively, the correction required is $\dfrac{50+40}{200} \times 66.3 = 66.3 \times 0.45 = 29.8$ cyds.

To find the correction for curvature in side-hill work when the transverse surface slope is regular.

Given : area; degree of curve; side distance; road-bed width; and width of excavation at road-bed (A; $C°$; d; b; w).

RULE 6 (FORMULA 24).

Enter table 18 with $d+b-w$ and take out the corresponding factor : multiply this by the product of A by $C°$, and the result is Q the correction in cubic yards, to be applied in all respects as in Rule 5.

Example.—Given $w = 17$; $b = 30$; $d = 51$; $h_1 = 24$; $R = 1600$; or $A = 204$; $C° = 3°.58$; $d+b-w = 64$.
$$64 \text{ table } 18 = 0.01379,$$
and $204 \times 3.58 \times 0.01379 = 10.1$ cyds.

If both intervals are 50 feet, the correction required is $\dfrac{50+50}{200} \times 10.1 = 10.1 \times 0.5 = 5$ cyds.

For correction for curvature when the transverse surface slope is broken, or for double-width thorough-cut, see page 24.

Rules 5 and 6 apply to *excavation* only. For *embankment* the correction is to be *added* or *subtracted* accordingly as the curve is *concave* or *convex* toward the higher ground.

MISCELLANEOUS EXAMPLES.

EXAMPLE 1.

(1) Stations.	(2) Distances.	(3) End Areas.	(4) Average Areas.	(5) Corr'n Areas.	(6) Corr'n sq. roots.	(7) Diff. sq. roots.	(8) Average Contents. cu. yds.	(9) Corr'n Contents. cu. yds.	(10) Prismoidal Contents. cu. yds.
0		0.0		0.0	0.0				
	80		30.0			7.7	88.9	29.3	59.6
a		60.0		{60.0 / 160.0}	{7.7 / 12.6}				
	60		96.2			2.6	213.8	2.5	211.3
1		132.5		232.5	15.2				
	100		190.9			3.5	707.0	7.6	699.4
3		249.2		349.2	18.7				
	100		280.9			1.6	1040.3	1.6	1038.7
5		312.7		412.7	20.3				
	100		466.6			6.5	1728.1	26.1	1702.0
7		620.5		720.5	26.8				
	100		682.6			2.3	2528.1	3.3	2524.8
9		744.8		844.8	29.1				
	100		864.9			3.8	3203.3	8.9	3194.4
11		985.0		1085.0	32.9				
	100		893.3			2.9	3308.5	5.2	3303.3
13		801.5		901.5	30.0				
	100		608.7			7.3	2254.4	32.9	2221.5
15		416.0		516.0	22.7				
	100		287.8			6.6	1065.9	26.9	1039.0
17		159.5		259.5	16.1				
	40		129.7			2.0	192.1	1.0	191.1
a		100.0		{200.0 / 100.0}	{14.1 / 10.0}				
	50		50			10.0	92.6	30.8	61.8
0		0.0		0.0	0.0				
							16423.0	−176.1	=16246.9

Example 1, as above, is of the railroad cut given in Morris's "Earthworks,"* pp. 47–54, with contents computed by Rules 1, 2, and 4, and the auxiliary tables of the present work. As here used, the areas are supposed to belong to sections which, when carried to the intersection of the side slopes in thorough-cut, are rendered sensibly similar, and the examples as here given are intended

* "Easy Rules for the Measurement of Earthworks by means of the Prismoidal Formula. By Ellwood Morris, C.E." Philadelphia: 1872.

to show only the comparative facility of arriving at the prismoidal contents by Mr. Morris's methods and those of the preceding rules when the above condition of similarity is fulfilled, and not to endorse the application of the method of "Roots and Squares" (or of the rules of this work) in cases where the hypothetical middle area materially differs from the actual one.*

Except by trial with the actual middle section and the prismoidal formula, it seems almost impossible in cases of dissimilar end sections to know when the application of the method of Roots and Squares, or of the preceding rules, begins to fail of practical correctness, but it may safely be assumed that if the ground is properly and sufficiently cross-sectioned, the results obtained by them will be practically the prismoidal contents.

The above tabulated example shows all the steps necessary in finding the prismoidal contents in cubic yards when the areas are given. Columns (1), (2), and (3) being written out, (4) is derived directly from (3) by averaging ; (5) from (3) by adding area of grade triangle in thorough-cut ; (6) from (5) by table 3 ; (7) from (6) by subtraction ; (8) from (4) by table 4 ; (9) from (7) by table 5 ; and (10) from (8) and (9) by subtraction.

Column (4) gives the average end areas throughout the cut, including the terminal pyramids, and the only break in the routine of adding the area of the grade triangle in column (5) is at the point where the cutting runs out on the lower side. At such points two areas have to be used, the one of earthwork *plus* the grade triangle, for computation of thorough-cut by Rule 1, and the other of earthwork alone, for the calculation of the pyramid or side-hill work into which the thorough-cut changes, and of which the computation of contents falls under Rule 2.

Column (8) gives the contents between each two stations roughed out by the common method of "average areas," column (9) the corresponding error, and column (10) the prismoidal contents, all in cubic yards.

It is not strictly necessary to write out all of the columns given above, but errors are so much more readily detected when all of the steps are shown, that ordinarily time and labor will be saved by adopting some system of tabulating similar to the above, both as regards the number of columns and the arrangement by which the figures referring to each two stations may be recorded on a line between them.

---

* See article on the application of the prismoidal formula, page 16.

The prismoidal contents in cubic yards between stations 1 and 17 are given by Mr. Morris as 15,721, and by the above computation as 15,723, whilst the contents of the whole cut given by him as 16,664 appear above as 16,247. The discrepancy is in the truncated portions of the cut outside of stations 1 and 17, which by some oversight he gives as 943, instead of 524 cubic yards.

The preceding example will now be computed by equivalent level heights and Rule 4. The data of level heights are supposed to be obtained from Trautwine's diagrams, as when such accuracy is required as renders the calculation of areas necessary, Rule 1, 2, or 3 should be used for the computation of contents.

EXAMPLE 2.

(1)	(2)	(3)	(4)	(5)	(6)	(7)	(8)
Stations.	Distances.	Eq. Level Heights.	Eq. Level Heights. Half-sum.	Eq. Level Heights. Difference.	End Heights. Contents.	Corr'n Contents. cu. yds.	Prismoidal Contents. cu. yds.
0		0.7					
	40		1.6	1.9	51	0	51
a		2.6					
	60		3.9	2.6	207	1	208
1		5.2					
	100		7.0	3.5	700	4	704
3		8.7					
	100		9.5	1.6	1038	1	1039
5		10.3					
	100		13.6	6.5	1692	13	1705
7		16.8					
	100		18.0	2.3	2533	2	2535
9		19.1					
	100		21.0	3.8	3189	5	3194
11		22.9					
	100		21.5	2.9	3305	3	3308
13		·20.0					
	100		16.4	7.3	2211	16	2227
15		12.7					
	100		9.4	6.6	1024	13	1037
17		6.1					
	40		5.1	2.0	190	0	190
a		4.1					
	25		2.6	3.1	54	1	55
0		1.0					
					16194	+ 59 =	16253

With equivalent level heights given, the above tabulated example shows all the steps required in finding the approximate prismoidal contents in cubic yards. Columns (1), (2), and (3) being written out, (4) is derived directly from (3) by averaging, and (5) from (3) by subtracting. The table of level cuttings for a base of 20 feet and slopes 1 to 1, from which column (6) should be taken, is not published in this volume, but its place may readily be supplied by adding 1. to each of the heights of column (3), and taking 70 from each of the corresponding quantities in table 12. Such remainders are the amounts in column (6). Column (7) is derived from (5) by table 14, and (8) from (6) and (7) by addition.

In ordinary ground sloping transversely, the area of earthwork of the terminal pyramid at the point where the centre height is nothing, is about one-fourth of the area of the section where the pyramid begins; and practically, as only small quantities are concerned, the equivalent level height corresponding may be taken as one-fourth of that corresponding to the area of the base of the pyramid.

The calculation of contents by equivalent level heights and tables is well suited for preliminary or approximate estimates, especially if, as in the present case, when the sum of the tenths of the end heights is uneven, the average is always taken as the tenth next *greater* than the actual half-sum.

The variation between the contents of the thorough-cut from 1 to 17, as given in Examples 1 and 2, is due to the fact that the equivalent level heights are carried out to tenths only. In the present case, at a height of 20 feet the increment is over two cubic yards for each 0.01 of a foot, and in embankment at the same height it is still greater. As in practice neither equivalent level heights nor those of the tables of level cuttings are carried out to hundredths, one cause of the greater accuracy of the previous method by Rules 1 and 2 is evident. It may be replied that errors as important are involved in the field work, the cross section stakes being set only approximately; but that an element of error should voluntarily be introduced into the calculations because another such already exists in the data, is a position that will not be contended for seriously.

Example 3.—In a cutting with road-bed width 16 feet, and opposite side slopes $\frac{1}{2}$ and $\frac{3}{4}$ to 1, the given areas of two consecutive cross sections with similar transverse surface lines and at a distance apart of 100 feet, are 100 and 1000 square feet respectively: required the prismoidal contents. Here the area of the grade triangle (table 2) is 102, and consequently the whole areas to intersection are 202 and 1102.

To find the correction numbers N and N'.

202 table 3 .... 14.2 = N
1102 table 3 .... 33.2 = N'

To find the contents in cubic yards.

$\dfrac{100+1000}{2}$ = 550 table 4 2037
14.2 ~ 33.2 = 19.0 table 5 .... —223
Contents for 100 feet .... 1814 cyds.

Test by Formula 9.

$\sqrt{202 \times 1102}$ = 472 = mean area to intersection.
$\left(\dfrac{202+1102+472}{3} - 102\right)\dfrac{100}{27} = \left(592-102\right)\dfrac{100}{27}$
= 490 table 4 .... 1815 cyds.

Example 4.—Given 100 and 1000 square feet respectively as the areas of two *similar* cross sections 100 feet apart, irrespective of shape or number of sides in perimeter : required the prismoidal contents.

To find the correction numbers n and n'.

100 table 3 .... 10.0 = n
1000 table 3 .... 31.6 = n'

To find the contents in cubic yards.

$\dfrac{100+1000}{2}$ = 550 table 4 2037
10.0 ~ 31.6 = 21.6 table 5 .... —288
Contents for 100 feet .... 1749 cyds.

Test by Formula 10.

$\sqrt{100 \times 1000}$ = 316 = mean area.
$\left(\dfrac{100+1000+316}{3}\right)\dfrac{100}{27} = 472\left(\dfrac{100}{27}\right)$
= 472 table 4 .... 1748 cyds.

Example 5.—At two stations 100 feet apart with base b = 14 feet, and side slopes s = 1¼ to 1, given the notes of the cross section at the first station, centre height C = 10.2, side heights h_1 and h_2 =

6.8 and 15.2, and side distances d and $d' = 17.2$ and 29.8 ; and at second station, centre height 38.6, side heights 28.6 and 53.0, and side distances 49.9 and 86.5.

Calculation of areas A and A', and correction numbers N and N'.

For the grade triangle corresponding to $b = 14$ and $\frac{s+s'}{2} = 1\frac{1}{2}$, the height table $1 = 4.67$, and the area table $2 = 33 = a$.

By Formula (1) and Rule 1.

Area $(A+a) = \frac{(10.2+4.67)(17.2+29.8)}{2} = 349$ table $3 = 18.7 =$ correction number N; and $349 - 33 = 316 = A$.

Area $(A'+a') = \frac{(38.6+4.67)(49.9+86.5)}{2} = 2951$ table $3 = 54.3$
$=$ correction number N'; and $2951 - 33 = 2918 = A'$.

Calculation of Contents.—Formula (18), Rule 1.

$\frac{316+2918}{2} = 1617$ table 4 .... 5989 cyds.

$18.7 \sim 54.3 = 35.6$ table 5 .... -782 "

Contents for 100 feet .... 5207 cyds.

Test by Formula 13.

From the preceding data the notes of the middle area would give centre height 24.4, and side distances 33.55 and 58.15 ; and by Formula (1)

$\frac{(24.4+4.67)(33.55+58.15)}{2} - 33 = 1333 - 33 = 1300 = M.$

by Formula (13) $\frac{317+2918+1300 \times 4}{6} \times \frac{100}{27} = 1406$ tab. $4 = 5207$ cyds.

To find the equivalent level heights.—(Rule 7.)
316 table 4....1170 table 10....10.6 equiv. lev. ht.
2918 table 4....10,807 table 10...39.7 "   "

Test by Trautwine's method, with level heights.
10.6 table 10............................1174
39.7 table 10...........................10,815
(25.15 table 10........4818.5)×4........19,274

6 | 31,263
Contents for 100 feet...........5,210.5 cyds.

By Formula (21), *Rule* (4), *with level heights.*

$\frac{10.6+39.7}{2} = 25.15$ table 10............4818.5

$10.6\sim39.7 = 29.1$ table 15.............+392.0

Contents for 100 feet...........5,210.5 cyds.

By Formula (20), *with end areas and level heights.*

$\frac{316+2918}{2} = 1617$ table 4................5989

$10.6\sim39.7 = 29.1$ table 17...............−784

Contents for 100 feet.............5205 cyds.

Approximation by Formula (20), *with centre heights of profile substituted for level heights.*

$\frac{316+2918}{2} = 1617$ table 4................5989

$10.2\sim38.6 = 28.4$ table 17...............−747

Approximate contents for 100 feet..5,242 cyds.

This approximation is for an extreme case, as in practice the difference between two consecutive centre heights is rarely as much as one-half of the difference above taken. In ordinary cases this approximation gives results very nearly correct.

It will be observed that by Trautwine's method, as given above, three quantities are taken from the tables, and that it involves an addition of three quantities, a multiplication, and a division; whilst by Rule 4, which with the same data gives the same result, the sum of two quantities taken from the tables is the required contents.

Example 6.—Correction of Contents for Curvature.—If the second cross section of Example 5 is at the middle one of three stations 100 feet apart, and all of them on a 6° curve which is concave toward the higher ground, the correction for curvature to be deducted at the station under consideration is obtained as follows by Rule 5 :

From the above C° = 6, and from the notes of Example 5, A+a = 2951, and $d\sim d' = 36.6$. But 36.6 table 18 = 0.007885; and Q = 2951 × 6 × 0.007885 = 139.6 cyds.

Test by Henck's Formula.

$C = \{\frac{1}{4}c(d-d')+\frac{1}{4}b(h-h')\} \times \frac{2}{3}(d+d') \sin D$, in which d and d' are side distances, h and h' side heights, c the centre height, and D the deflection angle; hence from the above and the notes of Example 5,

$$C = \left(\frac{38.6}{2} \times 36.6 + \frac{14 \times 24.4}{4}\right) \times \frac{2 \times 136.4}{3} \times 0.05234 = 3768.5 \text{ cu. feet}$$

$= 139.6$ cyds. In practice $d \sim d'$ is required to the nearest foot only.

REMARKS ON ESTIMATING CONTENTS.

PROFILE EARTHWORK.

In addition to the cross sections at the regular stations, others are necessary where changes begin in the character of the transverse surface slope, as well as at all points where the surface line of the profile changes its direction; and all of the formulæ and rules heretofore given for finding the contents suppose the solid to be between two consecutive cross sections taken at such points.

In passing from cutting into embankment, cross sections should always be taken at the two points on opposite sides of the road-bed where the cutting "runs out." This will obviate the necessity for staking out the "P.P." except with a zero point on the centre line, as, in addition to accurate data for calculation of the pyramids of cut and bank which lie between the two cross sections thus taken, two more zero points, one on each side of the road-bed, will be given. For like reasons, in passing from thorough into side hill cutting, the point on the lower side where the excavation runs out should be cross-sectioned.

Where the original quantities of excavation and embankment have been calculated, and the work is being done according to the slope-stakes and field-notes, probably the simplest method of obtaining the quantities moved in an unfinished cutting or embankment is to take the average heights above or below the road-bed at each of the several stations of that portion which has been worked upon, and then, with Formula (21), Rule 4, and tables, to calculate by these heights the quantities remaining to be done. The latter subtracted from the original quantities between the same stations will, of course, give the desired amount.

When the material lies in strata, a similar means may be used for determining the respective quantities of the different kinds of excavation. For example, a cutting may be composed of earth at top, loose rock below the earth, and solid rock at bottom: the amounts then calculated by the loose rock heights, and deducted from the original quantities giving the earth, and the solid rock similarly calculated and deducted from the amounts obtained by the loose rock heights giving the loose rock. When the necessary average heights have been obtained, the quantities corresponding may be found very rapidly by Rule 4 and the proper tables.

For approximate estimates, when the centre heights and transverse surface slopes only are given, the shortest method is to find the equivalent level heights by Trautwine's diagrams, and then take out the contents by Rule 4.

When the work is carried on irregularly, no general rules for ascertaining the true contents can be given. When the cross sections are very irregular and dissimilar, the best practical rule is to take them at very short intervals. This in all cases reduces the error in the calculation of contents to a minimum.

A very careful and thorough investigation of the mathematical methods of calculating irregular earthwork is given in the article on "Earthwork" in Henck's "Field-Book," and to that the theoretical reader is referred.

BORROW PITS.

For obtaining the contents of extensive borrow pits, the following will be found to be about as simple a method as is consistent with correctness. Before the excavation is commenced, lay off the surface in squares, rectangles, or triangles, small enough to be considered as plane surfaces, and take elevations with the Level at all of the corners. These elevations must be referred to a base which will be below the bottom of the borrow pit when the work is finished.

A plan of the ground as laid off should then be made, and the elevations above the base recorded on it at the corners. When an estimate of the quantities excavated is to be made during the progress of the work, the horizontal limits of the pit as then excavated should be taken, and inside of these limits the whole of the ground again divided into rectangles and triangles without reference to the former surface divisions, the elevations above the base plane again being taken at all corners, including those on the surface at the edges of the pit.

The original quantity inside of the pit limits and down to the base plane, taken as a series of truncated prisms, should then be calculated, and next the quantity remaining inside of the pit limits and above the base plane. The difference between those amounts gives the quantity excavated.

The advantage of using an independent method of dividing up the ground after the original surface has been removed is that it rarely happens that the best arrangement of these subdivisions for reducing to plane surfaces will agree accurately, either in size or position, with those originally taken on the ground surface. If, however, the same divisions can be taken in the bottom of the pit as originally on the surface, the differences of the elevations at each corner taken before and after the excavation is made will give the heights of the prisms, of which the contents may be obtained by a single calculation.

In order to prevent the necessity for recalculating the finished portions at each estimate, when any portion of the pit will not again be disturbed, its limits should be referenced on the ground and indicated on the plan, and its contents recorded separately.

RULES FOR VARIOUS USES OF TABLES.

To find the height of an equivalent level section.

*Given : areas, side slopes, and base.

RULE 7.

Enter table 4 with the given area, and take out the corresponding quantity: find the quantity nearest to this in the body of table of level cuttings with the given side slopes and base, and the index number corresponding is the equivalent level height to the nearest tenth.

* When centre heights and transverse surface slopes only are given, if $r =$ ratio to 1 of surface slope = cotangent of surface angle, and $s' = s$, then the equivalent level height $= h = \left(c + \dfrac{b}{2s}\right) \dfrac{r}{\sqrt{r^2 - s^2}} - \dfrac{b}{2s}$.

Example.—Given $a = 800$; $\frac{s+s'}{2} = 1\frac{1}{2}$; $b = 14$

800 table 4....2963 table 15....18.9 equiv. lev. ht.

To find the area corresponding to a level height, reverse the process of Rule 7.

To find the middle area of Rule 1.

Given: N, N', and a.

RULE 8.

Enter table 3 with $\frac{N+N'}{2}$, and take out the quantity corresponding; from this deduct a, and the remainder is the middle area.

From example 5, page 36, N = 18.7; N' = 54.3; and $a = 33$.

$$\frac{18.7 + 54.3}{2} = 36.5 \text{ table } 3 \ldots\ldots\ldots\ldots 1332$$

$$1332 - 33 = 1299 = M$$

To find the middle area of Rule 2.

Given: n and n'.

RULE 9.

Enter table 3 with $\frac{n+n'}{2}$, and the quantity corresponding is the middle area.

Example.—With similar end areas $4 \times 25 = 100$, and $8 \times 50 = 400$, the middle area is $6 \times 37.5 = 225$. Here $n = 10$ and $n' = 20$, and $\frac{n+n'}{2} = \frac{10+20}{2} = 15$ table $3 = 225 = M$.

To find the middle area of Rule 4.

Given: h and h'; $\frac{s+s'}{2}$; and b.

RULE 10.

Enter the table of level cuttings for the given side slopes and base with $\frac{h+h'}{2}$, and take out the corresponding quantity: find the quantity nearest to this in the body of table 4, and the index number corresponding is the middle area.

Example.—From example 5, page 36, $h = 10.6$ and $h' = 39.7$.

$$\frac{10.6 + 39.7}{2} = 25.15 \text{ table } 15 \ldots .4818 \text{ table } 4 \ldots .1301.$$

To extend the Correction Tables, general or special.

RULE 11.

When the difference of the correction numbers, or of the level heights, is too large to enter the table with, take one-half of it, and with this enter and take out the corresponding quantity, which multiplied by 4 gives the correction required for a length of 100 feet.

Examples.—In table 5 the correction corresponding to 32 is 632.1, which multiplied by 4 gives 2528.4, the correction corresponding to 64.

In table 17, the correction corresponding to 12.2 is 68.9, which multiplied by 4 gives 275.6, the correction corresponding to 24.4.

To find the special corrections for any given side slopes from the general correction table.

RULE 12.

Enter table 5 with $h \sim h'$, and take out the corresponding quantity; for the special *plus* corrections multiply this by the quarter-sum of the ratios of the side slopes $\left(\frac{s+s'}{4}\right)$; for the special *minus* correction multiply by the half-sum $\left(\frac{s+s'}{2}\right)$. The corrections so obtained are for = lengths of 100 feet.

Examples.—From table 5 the general minus correction corresponding to 39.4 is 958.2, and the plus correction for $\frac{s+s'}{2} = 1\frac{1}{2}$ is 958.2 × $\frac{3}{4}$ = 718.7 corresponding to 39.4 table 17. The minus correction for $\frac{s+s'}{2} = \frac{1}{2}$ is 958.2 × $\frac{1}{2}$ = 479.1 corresponding to 39.4 table 14. In like manner with $\frac{s+s'}{2} = \frac{1}{5}$ the plus correction for 39.4 = 958.2 × 0.1 = 95.8, table 8; and with $\frac{s+s'}{2} = 1$, the minus corrections, general and special, are the same, as are $N \sim N'$ and $h \sim h'$. (See table 5, and examples 1 and 2, pages 31 and 33.)

EXPLANATIONS OF TABLES.

Table 1 is for obtaining the height of the grade triangle. To use it, find the half-sum of the ratios of the given side slopes at the top, and the number vertically below, and on the same line with the given road-bed width in the left column, is the height required. Thus with $b = 16$ and $\frac{s+s'}{2} = \frac{5}{8}$ the height corresponding is 12.8.

Table 2 contains the area of the same triangle. It is used with the same data and entered in the same way. Thus with $b = 18$ and $\frac{s+s'}{2} = \frac{1}{4}$ the area corresponding $= a = 162$.

Table 3 gives square roots to tenths, or correction numbers of areas. To use it, find in the body of the table the number nearest to that which expresses the area under consideration, and the figures on the same horizontal line in the left column are the whole numbers, and that immediately above it, at the top of the table, the tenths of the correction number required. Thus if the area to intersection of side slopes is 2,000, the correction number N is 44.7 : if one of similar end areas is 230, the correction number n is 15.2.

Table 4 is for finding the contents for 100 feet corresponding to a given area. The left column contains the tens, and the top the units, of the area. In the body of the table are the corresponding contents in cubic yards for lengths of 100 feet. In the short table of two lines prefixed, the contents corresponding to the tenths of the area are given, and these when required are to be added to the contents taken from the main table. Thus the contents corresponding to the area 1872.7 are $6933.3 + 2.6 = 6935.9$ cubic yards.

Table 5 is for obtaining the corrections for computations by average areas. The arithmetical difference between the correction numbers is to be found in whole numbers and tenths respectively, in the left column and at the top of the table, and the number corresponding in the body of the table is the correction in cubic yards for a length of 100 feet. Thus if the difference of the correction numbers is 28.3, the correction corresponding is 494.4 cyds. This correction is always to be subtracted.

The Tables of Level Cuttings for special side slopes and road-bed widths give the cubic yards for lengths of 100 feet corresponding to the different heights, of which the whole numbers are in the left column and the tenths at top.

The special tables of *plus* corrections give the correction for computation by averaging equivalent level heights. The differences of the end heights in feet and tenths respectively are in the left column and at top, and the corresponding corrections for lengths of 100 feet in the body of the table. Care must be taken to use the correction table with the half sum of the side slopes the same as that of the table of level cuttings of which the contents are to be corrected.

The special tables of *minus* corrections give the corrections for average areas when entered with the heights of equivalent level sections. The side slopes of the table must be the same as those of the end sections, between which the contents are to be corrected.

When the tables of *minus* corrections for special slopes are entered with the differences of the centre heights of the profile instead of those of the equivalent level heights, in ordinary ground a close approximation to the true correction is obtained.

For the special *plus* correction tables the half-sum of the side slopes is indicated at the *top*. For the special *minus* corrections the slopes are indicated at the *bottom* of the same tables.

Table 18 contains factors for calculation of the corrections for curvature. Its use is explained in Rules 5 and 6.

TABLE No. 1.

Roadbed Width in Left Column; half-sum of ratios of Side Slopes at Top; Height of Grade Triangle in body of Table.

Feet	$\frac{1}{8}$	$\frac{1}{4}$	$\frac{3}{8}$	$\frac{1}{2}$	$\frac{5}{8}$	$\frac{3}{4}$	$\frac{7}{8}$	1	$1\frac{1}{8}$	$1\frac{1}{4}$	$1\frac{3}{8}$	$1\frac{1}{2}$	2
10	25	20	13.3	10	8.0	6.7	5.7	5	4.4	4.0	3.6	3.3	2.5
12	30	24	16.0	12	9.6	8.0	6.9	6	5.3	4.8	4.4	4.0	3 0
14	35	28	18.7	14	11.2	9.3	8.0	7	6.2	5.6	5.1	4.7	3.5
16	40	32	21.3	16	12.8	10.7	9.1	8	7.1	6.4	5.8	5.3	4.0
18	45	36	24.0	18	14.4	12.0	10.3	9	8.0	7.2	6.5	6.0	4.5
20	50	40	26.7	20	16.0	13.3	11.4	10	8.9	8.0	7.3	6.7	5.0
22	55	44	29.3	22	17.6	14.7	12.6	11	9.8	8.8	8.0	7.3	5.5
24	60	48	32.0	24	19.2	16.0	13.7	12	10.7	9.6	8 7	8.0	6.0
26	65	52	34.7	26	20.8	17.3	14.9	13	11.6	10.4	9.5	8.7	6.5
28	70	56	37.3	28	22.4	18.7	16.0	14	12.4	11.2	10.2	9.3	7.0
30	75	60	40.0	30	24.0	20.0	17.1	15	13.3	12.0	10.9	10.0	7.5
	$\frac{1}{8}$	$\frac{1}{4}$	$\frac{3}{8}$	$\frac{1}{2}$	$\frac{5}{8}$	$\frac{3}{4}$	$\frac{7}{8}$	1	$1\frac{1}{8}$	$1\frac{1}{4}$	$1\frac{3}{8}$	$1\frac{1}{2}$	2

TABLE No. 2.

Roadbed Width in Left Column; half-sum of ratios of Side Slopes at Top; Area of Grade Triangle in body of Table.

Feet	$\frac{1}{8}$	$\frac{1}{4}$	$\frac{3}{8}$	$\frac{1}{2}$	$\frac{5}{8}$	$\frac{3}{4}$	$\frac{7}{8}$	1	$1\frac{1}{8}$	$1\frac{1}{4}$	$1\frac{3}{8}$	$1\frac{1}{2}$	2
10	125	100	66.7	50	40.0	33.3	28.6	25	22.2	20.0	18.2	16.7	12.5
12	180	144	96.0	72	57.6	48.0	41.1	36	32.0	28.8	26.2	24.0	18.0
14	245	196	130.7	98	78.4	65.3	56.0	49	43.5	38.2	35.6	32.7	24.5
16	320	256	170.7	128	102.4	85.3	73.1	64	56.9	51.2	46.6	42.7	32.0
18	405	324	216.0	162	129.6	108.0	92.6	81	72.0	64.8	58.9	54.0	40.5
20	500	400	266.7	200	160.0	133.3	114.3	100	88.9	80.0	72.7	66.7	50.0
22	605	484	322.7	242	193.6	161.3	138.3	121	107.5	96.8	88.0	80.7	60.5
24	720	576	384.0	288	230.4	192.0	164.6	144	128.0	115.2	104.7	96.0	72.0
26	845	676	450.7	338	270.4	225.3	193.1	169	150.2	135.2	122.9	112.7	84.5
28	980	784	522.7	392	313.6	261.3	224.0	196	174.2	156.8	142.6	130.7	98.0
30	1125	900	600.0	450	360.0	300.0	257.1	225	200.0	180.0	163.6	150.0	112.5
	$\frac{1}{8}$	$\frac{1}{4}$	$\frac{3}{8}$	$\frac{1}{2}$	$\frac{5}{8}$	$\frac{3}{4}$	$\frac{7}{8}$	1	$1\frac{1}{8}$	$1\frac{1}{4}$	$1\frac{3}{8}$	$1\frac{1}{2}$	2

TABLE No. 3.

Areas in body of Table; Correction Nos., in feet and tenths, in left column and at top.

Feet	0	.1	.2	.3	.4	.5	.6	.7	.8	.9	Diff. for 0.05
0	0	0.0	0.0	0.1	0.2	0.3	0.4	0.5	0.6	0.8	0.05
1	1	1.2	1.4	1.7	2.	2.3	2.6	2.9	3.2	3.6	0.2
2	4	4.4	4.8	5.3	5.8	6.3	6.8	7.3	7.8	8.4	0.3
3	9	9.6	10.2	10.9	11.6	12.3	13.	13.7	14.4	15.2	0.4
4	16	16.8	17.6	18.5	19.4	20.3	21.2	22.1	23.	24.	0.5
5	25	26.	27.	28.1	29.2	30.3	31.4	32.5	33.6	34.8	0.6
6	36	37.2	38.4	39.7	41.	42.3	43.6	44.9	46.2	47.6	0.7
7	49	50.4	51.8	53.3	54.8	56.3	57.8	59.3	60.8	62.4	0.8
8	64	65.6	67.2	68.9	70.6	72.3	74.	75.7	77.4	79.2	0.9
9	81	82.8	84.6	86.5	88.4	90.3	92.2	94.1	96.	98.	1.
10	100	102.	104.	106.1	108.2	110.3	112.4	114.5	116.6	118.8	1.1
11	121	123.2	125.4	127.7	130.	132.3	134.6	136.9	139.2	141.6	1.2
12	144	146.4	148.8	151.3	153.8	156.3	158.8	161.3	163.8	166.4	1.3
13	169	171.6	174.2	176.9	179.6	182.3	185.	187.7	190.4	193.2	1.4
14	196	198.8	201.6	204.5	207.4	210.3	213.2	216.1	219.	222.	1.5
15	225	228.	231.	234.1	237.2	240.3	243.4	246.5	249.6	252.8	1.6
16	256	259.2	262.4	265.7	269.	272.3	275.6	278.9	282.2	285.6	1.7
17	289	292.4	295.8	299.3	302.8	306.3	309.8	313.3	316.8	320.4	1.8
18	324	327.6	331.2	334.9	338.6	342.3	346.	349.7	353.4	357.2	1.9
19	361	364.8	368.6	372.5	376.4	380.3	384.2	388.1	392.	396.	2.
20	400	404.	408.	412.1	416.2	420.3	424.4	428.5	432.6	436.8	2.1
21	441	445.2	449.4	453.7	458.	462.3	466.6	470.9	475.2	479.6	2.2
22	484	488.4	492.8	497.3	501.8	506.3	510.8	515.3	519.8	524.4	2.3
23	529	533.6	538.2	542.9	547.6	552.3	557.	561.7	566.4	571.2	2.4
24	576	580.8	585.6	590.5	595.4	600.3	605.2	610.1	615.	620.	2.5
25	625	630.	635.	640.1	645.2	650.3	655.4	660.5	665.6	670.3	2.6
26	676	681.2	686.4	691.7	697.	702.3	707.6	712.9	718.2	723.6	2.7
27	729	734.4	739.8	745.3	750.8	756.3	761.8	767.3	772.8	778.4	2.8
28	784	789.6	795.2	800.9	806.6	812.3	818.	823.7	829.4	835.2	2.9
29	841	846.8	852.6	858.5	864.4	870.3	876.2	882.1	888.	894.	3.0
30	900	906.	912.	918.1	924.2	930.3	936.4	942.5	948.6	954.8	3.1
31	961	967.2	973.4	979.7	986	992.3	998.6	1005	1011	1018	3.2
32	1024	1030	1037	1043	1050	1056	1063	1069	1076	1082	3.3
33	1089	1096	1102	1109	1116	1122	1129	1136	1142	1149	3.5
34	1156	1163	1170	1176	1183	1190	1197	1204	1211	1218	3.6
35	1225	1232	1239	1246	1253	1260	1267	1274	1282	1289	3.6
36	1296	1303	1310	1318	1325	1332	1340	1347	1354	1362	3.7
37	1369	1376	1384	1391	1399	1406	1414	1421	1429	1436	3.8
38	1444	1452	1459	1467	1475	1482	1490	1498	1505	1513	3.9
39	1521	1529	1537	1544	1552	1560	1568	1576	1584	1592	4.0
40	1600	1608	1616	1624	1632	1640	1648	1656	1665	1673	4.1
41	1681	1689	1697	1706	1714	1722	1731	1739	1747	1756	4.2
42	1764	1772	1781	1789	1798	1806	1815	1823	1832	1840	4.2
43	1849	1858	1866	1875	1884	1892	1901	1910	1918	1927	4.3
44	1936	1945	1954	1962	1971	1980	1989	1998	2007	2016	4.4
45	2025	2034	2043	2052	2061	2070	2079	2088	2098	2107	4.5
46	2116	2125	2134	2144	2153	2162	2172	2181	2190	2200	4.7
47	2209	2218	2228	2237	2247	2256	2266	2275	2285	2294	4.8
48	2304	2314	2323	2333	2343	2352	2362	2372	2381	2391	4.8
49	2401	2411	2421	2430	2440	2450	2460	2470	2480	2490	5.0
50	2500	2510	2520	2530	2540	2550	2560	2570	2581	2591	5.0
	0	.1	.2	.3	.4	.5	.6	.7	.8	.9	

TABLE No. 3—CONCLUDED.

Areas in body of Table; Correction Nos., in feet and tenths, in left column and at top.

Feet	0	.1	.2	.3	.4	.5	.6	.7	.8	.9	Diff. for 0.05
51	2601	2611	2621	2632	2642	2652	2663	2673	2683	2694	5.2
52	2704	2714	2725	2735	2746	2756	2767	2777	2788	2798	5.2
53	2809	2820	2830	2841	2852	2862	2873	2884	2894	2905	5.3
54	2916	2927	2938	2948	2959	2970	2981	2992	3003	3014	5.4
55	3025	3036	3047	3058	3069	3080	3091	3102	3114	3125	5.5
56	3136	3147	3158	3170	3181	3192	3204	3215	3226	3238	5.7
57	3249	3260	3272	3283	3295	3306	3318	3329	3341	3352	5.7
58	3364	3376	3387	3399	3411	3422	3434	3446	3457	3469	5.8
59	3481	3493	3505	3516	3528	3540	3552	3564	3576	3588	5.9
60	3600	3612	3624	3636	3648	3660	3672	3684	3697	3709	6.0
61	3721	3733	3745	3758	3770	3782	3795	3807	3819	3832	6.2
62	3844	3856	3869	3881	3894	3906	3919	3931	3944	3956	6.2
63	3969	3982	3994	4007	4020	4032	4045	4058	4070	4083	6.3
64	4096	4109	4122	4134	4147	4160	4173	4186	4199	4212	6.4
65	4225	4238	4251	4264	4277	4290	4303	4316	4330	4343	6.5
66	4356	4369	4382	4396	4409	4422	4436	4449	4462	4476	6.7
67	4489	4502	4516	4529	4543	4556	4570	4583	4597	4610	6.7
68	4624	4638	4651	4665	4679	4692	4706	4720	4733	4747	6.8
69	4761	4775	4789	4802	4816	4830	4844	4858	4872	4886	6.9
70	4900	4914	4928	4942	4956	4970	4984	4998	5013	5027	7.0
71	5041	5055	5069	5084	5098	5112	5127	5141	5155	5170	7.2
72	5184	5198	5213	5227	5242	5256	5271	5285	5300	5314	7.2
73	5329	5344	5358	5373	5388	5402	5417	5432	5446	5461	7.3
74	5476	5491	5506	5520	5535	5550	5565	5580	5595	5610	7.4
75	5625	5640	5655	5670	5685	5700	5715	5730	5746	5761	7.5
76	5776	5791	5806	5822	5837	5852	5868	5883	5898	5914	7.7
77	5929	5944	5960	5975	5991	6006	6022	6037	6053	6068	7.7
78	6084	6100	6115	6131	6147	6162	6178	6194	6209	6225	7.8
79	6241	6257	6273	6288	6304	6320	6336	6352	6368	6384	7.9
80	6400	6416	6432	6448	6464	6480	6496	6512	6529	6545	8.0
81	6561	6577	6593	6610	6626	6642	6659	6675	6691	6708	8.2
82	6724	6740	6757	6773	6790	6806	6823	6839	6856	6872	8.2
83	6889	6906	6922	6939	6956	6972	6989	7006	7022	7039	8.3
84	7056	7073	7090	7106	7123	7140	7157	7174	7191	7208	8.4
85	7225	7242	7259	7276	7293	7310	7327	7344	7362	7379	8.5
86	7396	7413	7430	7448	7465	7482	7500	7517	7534	7552	8.6
87	7569	7586	7604	7621	7639	7656	7674	7691	7709	7726	8.7
88	7744	7762	7779	7797	7815	7832	7850	7868	7885	7903	8.8
89	7921	7939	7957	7974	7992	8010	8028	8046	8064	8082	8.9
90	8100	8118	8136	8154	8172	8190	8208	8226	8245	8263	9.0
91	8281	8299	8317	8336	8354	8372	8391	8409	8427	8446	9.2
92	8464	8482	8501	8519	8538	8556	8575	8593	8612	8630	9.2
93	8649	8668	8686	8705	8724	8742	8761	8780	8798	8817	9.3
94	8836	8855	8874	8892	8911	8930	8949	8968	8987	9006	9.4
95	9025	9044	9063	9082	9101	9120	9139	9158	9178	9197	9.5
96	9216	9235	9254	9274	9293	9312	9332	9351	9370	9390	9.6
97	9409	9428	9448	9467	9487	9506	9526	9545	9565	9584	9.7
98	9604	9624	9643	9663	9683	9702	9722	9742	9761	9781	9.8
99	9801	9821	9841	9860	9880	9900	9920	9940	9960	9980	9.9
100	10000	10020	10040	10060	10080	10100	10120	10140	10161	10181	10.0
	0	.1	.2	.3	.4	.5	.6	.7	.8	.9	

TABLE No. 4.

Areas	0.1	0.2	0.25	0.3	0.4	0.5	0.6	0.7	0.75	0.8	0.9
Contents	0.4	0.7	0.9	1.1	1.5	1.9	2.2	2.6	2.8	3.0	3.3

Areas : Tens in left Column and Units at top. Contents for 100 feet in cubic yards in body of Table.

Feet	0.0	1.0	2.0	3.0	4.0	5.0	6.0	7.0	8.0	9.0
0	0.0	3.7	7.4	11.1	14.8	18.5	22.2	25.9	29.6	33.3
1	37.	40.7	44.4	48.1	51.9	55.6	59.3	63.	66.7	70.4
2	74.1	77.8	81.5	85.2	88.9	92.6	96.3	100.	103.7	107.4
3	111.1	114.8	118.5	122.2	125.9	129.6	133.3	137.	140.7	144.4
4	148.1	151.9	155.6	159.3	163.	166.7	170.4	174.1	177.8	181.5
5	185.2	188.9	192.6	196.3	200.	203.7	207.4	211.1	214.8	218.5
6	222.2	225.9	229.6	233.3	237.	240.7	244.4	248.1	251.9	255.6
7	259.3	263.	266.7	270.4	274.1	277.8	281.5	285.2	288.9	292.6
8	296.3	300.	303.7	307.4	311.1	314.8	318.5	322.2	325.9	329.6
9	333.3	337.	340.7	344.4	348.1	351.9	355.6	359.3	363.	366.7
10	370.4	374.1	377.8	381.5	385.2	388.9	392.6	396.3	400.	403.7
11	407.4	411.1	414.8	418.5	422.2	425.9	429.6	433.3	437.	440.7
12	444.4	448.1	451.9	455.6	459.3	463.	466.7	470.4	474.1	477.8
13	481.5	485.2	488.9	492.6	496.3	500.	503.7	507.4	511.1	514.8
14	518.5	522.2	525.9	529.6	533.3	537.	540.7	544.4	548.1	551.9
15	555.6	559.3	563.	566.7	570.4	574.1	577.8	581.5	585.2	588.9
16	592.6	596.3	600.	603.7	607.4	611.1	614.8	618.5	622.2	625.9
17	629.6	633.3	637.	640.7	644.4	648.1	651.9	655.6	659.3	663.
18	666.7	670.4	674.1	677.8	681.5	685.2	688.9	692.6	696.3	700.
19	703.7	707.4	711.1	714.8	718.5	722.2	725.9	729.6	733.3	737.
20	740.7	744.4	748.1	751.9	755.6	759.3	763.	766.7	770.4	774.1
21	777.8	781.5	785.2	788.9	792.6	796.3	800.	803.7	807.4	811.1
22	814.8	818.5	822.2	825.9	829.6	833.3	837.	840.7	844.4	848.1
23	851.9	855.6	859.3	863.	866.7	870.4	874.1	877.8	881.5	885.2
24	888.9	892.6	896.3	900.	903.7	907.4	911.1	914.8	918.5	922.2
25	925.9	929.6	933.3	937.	940.7	944.4	948.1	951.9	955.6	959.3
26	963.	966.7	970.4	974.1	977.8	981.5	985.2	988.9	992.6	996.3
27	1000.	1003.7	1007.4	1011.1	1014.8	1018.5	1022.2	1025.9	1029.6	1033.3
28	1037.	1040.7	1044.4	1048.1	1051.9	1055.6	1059.3	1063.	1066.7	1070.4
29	1074.1	1077.8	1081.5	1085.2	1088.9	1092.6	1096.3	1100.	1103.7	1107.4
30	1111.1	1114.8	1118.5	1122.2	1125.9	1129.6	1133.3	1137.	1140.7	1144.4
31	1148.1	1151.9	1155.6	1159.3	1163.	1166.7	1170.4	1174.1	1177.8	1181.5
32	1185.2	1188.9	1192.6	1196.3	1200.	1203.7	1207.4	1211.1	1214.8	1218.5
33	1222.2	1225.9	1229.6	1233.3	1237.	1240.7	1244.4	1248.1	1251.9	1255.6
34	1259.3	1263.	1266.7	1270.4	1274.1	1277.8	1281.5	1285.2	1288.9	1292.6
35	1296.3	1300.	1303.7	1307.4	1311.1	1314.8	1318.5	1322.2	1325.9	1329.6
36	1333.3	1337.	1340.7	1344.4	1348.1	1351.9	1355.6	1359.3	1363.	1366.7
37	1370.4	1374.1	1377.8	1381.5	1385.2	1388.9	1392.6	1396.3	1400.	1403.7
38	1407.4	1411.1	1414.8	1418.5	1422.2	1425.9	1429.6	1433.3	1437.	1440.7
39	1444.4	1448.1	1451.9	1455.6	1459.3	1463.	1466.7	1470.4	1474.1	1477.8
40	1481.5	1485.2	1488.9	1492.6	1496.3	1500.	1503.7	1507.4	1511.1	1514.8
41	1518.5	1522.2	1525.9	1529.6	1533.3	1537.	1540.7	1544.4	1548.1	1551.9
42	1555.6	1559.3	1563.	1566.7	1570.4	1574.1	1577.8	1581.5	1585.2	1588.9
43	1592.6	1596.3	1600.	1603.7	1607.4	1611.1	1614.8	1618.5	1622.2	1625.9
44	1629.6	1633.3	1637.	1640.7	1644.4	1648.1	1651.9	1655.6	1659.3	1663.
45	1666.7	1670.4	1674.1	1677.8	1681.5	1685.2	1688.9	1692.6	1696.3	1700.
46	1703.7	1707.4	1711.1	1714.8	1718.5	1722.2	1725.9	1729.6	1733.3	1737.
47	1740.7	1744.4	1748.1	1751.9	1755.6	1759.3	1763.	1766.7	1770.4	1774.1
48	1777.8	1781.5	1785.2	1788.9	1792.6	1796.3	1800.	1803.7	1807.4	1811.1
49	1814.8	1818.5	1822.2	1825.9	1829.6	1833.3	1837.	1840.7	1844.4	1848.1
50	1851.9	1855.6	1859.3	1863.	1866.7	1870.4	1874.1	1877.8	1881.5	1885.2
	0.	1.	2.	3.	4.	5.	6.	7.	8.	9.

TABLE No. 4—CONTINUED.

Areas	0.1	0.2	0.25	0.3	0.4	0.5	0.6	0.7	0.75	0.8	0.9
Contents	0.4	0.7	0.9	1.1	1.5	1.9	2.2	2.6	2.8	3.0	3.3

Areas: Tens in left Column and Units at top. Contents for 100 feet in cubic yards in body of Table.

Feet	0.	1.	2.	3.	4.	5.	6.	7.	8.	9.
51	1888.9	1892.6	1896.3	1900.	1903.7	1907.4	1911.1	1914.8	1918.5	1922.2
52	1925.9	1929.6	1933.3	1937.	1940.7	1944.4	1948.1	1951.9	1955.6	1959.3
53	1963.	1966.7	1970.4	1974.1	1977.8	1981.5	1985.2	1988.9	1992.6	1996.3
54	2000.	2003.7	2007.4	2011.1	2014.8	2018.5	2022.2	2025.9	2029.6	2033.3
55	2037.	2040.7	2044.4	2048.1	2051.9	2055.6	2059.3	2063.	2066.7	2070.4
56	2074.1	2077.8	2081.5	2085.2	2088.9	2092.6	2096.3	2100.	2103.7	2107.4
57	2111.1	2114.8	2118.5	2122.2	2125.9	2129.6	2133.3	2137.	2140.7	2144.4
58	2148.1	2151.9	2155.6	2159.3	2163.	2166.7	2170.4	2174.1	2177.8	2181.5
59	2185.2	2188.9	2192.6	2196.3	2200.	2203.7	2207.4	2211.1	2214.8	2218.5
60	2222.2	2225.9	2229.6	2233.3	2237.	2240.7	2244.4	2248.1	2251.9	2255.6
61	2259.3	2263.	2266.7	2270.4	2274.1	2277.8	2281.5	2285.2	2288.9	2292.6
62	2296.3	2300.	2303.7	2307.4	2311.1	2314.8	2318.5	2322.2	2325.9	2329.6
63	2333.3	2337.	2340.7	2344.4	2348.1	2351.9	2355.6	2359.3	2363.	2366.7
64	2370.4	2374.1	2377.8	2381.5	2385.2	2388.9	2392.6	2396.3	2400.	2403.7
65	2407.4	2411.1	2414.8	2418.5	2422.2	2425.9	2429.6	2433.3	2437.	2440.7
66	2444.4	2448.1	2451.9	2455.6	2459.3	2463.	2466.7	2470.4	2474.1	2477.8
67	2481.5	2485.2	2488.9	2492.6	2496.3	2500.	2503.7	2507.4	2511.1	2514.8
68	2518.5	2522.2	2525.9	2529.6	2533.3	2537.	2540.7	2544.4	2548.1	2551.9
69	2555.6	2559.3	2563.	2566.7	2570.4	2574.1	2577.8	2581.5	2585.2	2588.9
70	2592.6	2596.3	2600.	2603.7	2607.4	2611.1	2614.8	2618.5	2622.2	2625.9
71	2629.6	2633.3	2637.	2640.7	2644.4	2648.1	2651.9	2655.6	2659.3	2663.
72	2666.7	2670.4	2674.1	2677.8	2681.5	2685.2	2688.9	2692.6	2696.3	2700.
73	2703.7	2707.4	2711.1	2714.8	2718.5	2722.2	2725.9	2729.6	2733.3	2737.
74	2740.7	2744.4	2748.1	2751.9	2755.6	2759.3	2763.	2766.7	2770.4	2774.1
75	2777.8	2781.5	2785.2	2788.9	2792.6	2796.3	2800.	2803.7	2807.4	2811.1
76	2814.8	2818.5	2822.2	2825.9	2829.6	2833.3	2837.	2840.7	2844.4	2848.1
77	2851.9	2855.6	2859.3	2863.	2866.7	2870.4	2874.1	2877.8	2881.5	2885.2
78	2888.9	2892.6	2896.3	2900.	2903.7	2907.4	2911.1	2914.8	2918.5	2922.2
79	2925.9	2929.6	2933.3	2937.	2940.7	2944.4	2948.1	2951.9	2955.6	2959.3
80	2963.	2966.7	2970.4	2974.1	2977.8	2981.5	2985.2	2988.9	2992.6	2996.3
81	3000.	3003.7	3007.4	3011.1	3014.8	3018.5	3022.2	3025.9	3029.6	3033.3
82	3037.	3040.7	3044.4	3048.1	3051.9	3055.6	3059.3	3063.	3066.7	3070.4
83	3074.1	3077.8	3081.5	3085.2	3088.9	3092.6	3096.3	3100.	3103.7	3107.4
84	3111.1	3114.8	3118.5	3122.2	3125.9	3129.6	3133.3	3137.	3140.7	3144.4
85	3148.1	3151.9	3155.6	3159.3	3163.	3166.7	3170.4	3174.1	3177.8	3181.5
86	3185.2	3188.9	3192.6	3196.3	3200.	3203.7	3207.4	3211.1	3214.8	3218.5
87	3222.2	3225.9	3229.6	3233.3	3237.	3240.7	3244.4	3248.1	3251.9	3255.6
88	3259.3	3263.	3266.7	3270.4	3274.1	3277.8	3281.5	3285.2	3288.9	3292.6
89	3296.3	3300.	3303.7	3307.4	3311.1	3314.8	3318.5	3322.2	3325.9	3329.6
90	3333.3	3337.	3340.7	3344.4	3348.1	3351.9	3355.6	3359.3	3363.	3366.7
91	3370.4	3374.1	3377.8	3381.5	3385.2	3388.9	3392.6	3396.3	3400.	3403.7
92	3407.4	3411.1	3414.8	3418.5	3422.2	3425.9	3429.6	3433.3	3437.	3440.7
93	3444.4	3448.1	3451.9	3455.6	3459.3	3463.	3466.7	3470.4	3474.1	3477.8
94	3481.5	3485.2	3488.9	3492.6	3496.3	3500.	3503.7	3507.4	3511.1	3514.8
95	3518.5	3522.2	3525.9	3529.6	3533.3	3537.	3540.7	3544.4	3548.1	3551.9
96	3555.6	3559.3	3563.	3566.7	3570.4	3574.1	3577.8	3581.5	3585.2	3588.9
97	3592.6	3596.3	3600.	3603.7	3607.4	3611.1	3614.8	3618.5	3622.2	3625.9
98	3629.6	3633.3	3637.	3640.7	3644.4	3648.1	3651.9	3655.6	3659.3	3663.
99	3666.7	3670.4	3674.1	3677.8	3681.5	3685.2	3688.9	3692.6	3696.3	3700.
100	3703.7	3707.4	3711.1	3714.8	3718.5	3722.2	3725.9	3729.6	3733.3	3737.
	0.	1.	2.	3.	4.	5.	6.	7.	8.	9.

TABLE No. 4—Continued.

Areas	0.1	0.2	0.25	0.3	0.4	0.5	0.6	0.7	0.75	0.8	0.9
Contents	0.4	0.7	0.9	1.1	1.5	1.9	2 2	2.6	2.8	3.0	3.3

Areas: Tens in left Column and Units at top.   Contents for 100 feet in cubic yards in body of Table.

Feet	0.	1.	2.	3.	4.	5.	6.	7.	8.	9.
101	3740.7	3744.4	3748.1	3751.9	3755.6	3759.3	3763.	3766.7	3770.4	3774.1
102	3777.8	3781.5	3785.2	3788.9	3792.6	3796.3	3800.	3803.7	3807.4	3811.1
103	3814.8	3818.5	3822.2	3825.9	3829.6	3833.3	3837.	3840.7	3844.4	3848.1
104	3851.9	3855.6	3859.3	3863.	3866.7	3870.4	3874.1	3877.8	3881.5	3885.2
105	3888.9	3892.6	3896.3	3900.	3903.7	3907.4	3911.1	3914.8	3918.5	3922.2
106	3925.9	3929.6	3933.3	3937.	3940.7	3944.4	3948.1	3951.9	3955.6	3959.3
107	3963.	3966.7	3970.4	3974.1	3977.8	3981.5	3985.2	3988.9	3992.6	3996.3
108	4000.	4003.7	4007.4	4011.1	4014.8	4018.5	4022.2	4025.9	4029.6	4033.3
109	4037.	4040.7	4044.4	4048.1	4051.9	4055.6	4059.3	4063.	4066.7	4070.4
110	4074.1	4077.8	4081.5	4085.2	4088.9	4092.6	4096.3	4100.	4103.7	4107.4
111	4111.1	4114.8	4118.5	4122.2	4125.9	4129.6	4133.3	4137.	4140.7	4144.4
112	4148.1	4151.9	4155.6	4159.3	4163.	4166.7	4170.4	4174.1	4177.8	4181.5
113	4185.2	4188.9	4192.6	4196.3	4200.	4203.7	4207.4	4211.1	4214.8	4218.5
114	4222.2	4225.9	4229.6	4233.3	4237.	4240.7	4244.4	4248.1	4251.9	4255.6
115	4259.3	4263.	4266.7	4270.4	4274.1	4277.8	4281.5	4285.2	4288.9	4292.6
116	4296.3	4300.	4303.7	4307.4	4311.1	4314.8	4318.5	4322.2	4325.9	4329.6
117	4333.3	4337.	4340.7	4344.4	4348.1	4351.9	4355.6	4359.3	4363.	4366.7
118	4370.4	4374.1	4377.8	4381.5	4385.2	4388.9	4392.6	4396.3	4400.	4403.7
119	4407.4	4411.1	4414.8	4418.5	4422.2	4425.9	4429.6	4433.3	4437.	4440.7
120	4444.4	4448.1	4451.9	4455.6	4459.3	4463.	4466.7	4470.4	4474.1	4477.8
121	4481.5	4485.2	4488.9	4492.6	4496.3	4500.	4503.7	4507.4	4511.1	4514.8
122	4518.5	4522.2	4525.9	4529.6	4533.3	4537.	4540.7	4544.4	4548.1	4551.9
123	4555.6	4559.3	4563.	4566.7	4570.4	4574.1	4577.8	4581.5	4585.2	4588.9
124	4592.6	4596.3	4600.	4603.7	4607.4	4611.1	4614.8	4618.5	4622.2	4625.9
125	4629.6	4633.3	4637.	4640.7	4644.4	4648.1	4651.9	4655.6	4659.3	4663.
126	4666.7	4670.4	4674.1	4677.8	4681.5	4685.2	4688.9	4692.6	4696.3	4700.
127	4703.7	4707.4	4711.1	4714.8	4718.5	4722.2	4725.9	4729.6	4733.3	4737.
128	4740.7	4744.4	4748.1	4751.9	4755.6	4759.3	4763.	4766.7	4770.4	4774.1
129	4777.8	4781.5	4785.2	4788.9	4792.6	4796.3	4800.	4803.7	4807.4	4811.1
130	4814.8	4818.5	4822.2	4825.9	4829.6	4833.3	4837.	4840.7	4844.4	4848.1
131	4851.9	4855.6	4859.3	4863.	4866.7	4870.4	4874.1	4877.8	4881.5	4885.2
132	4888.9	4892.6	4896.3	4900.	4903.7	4907.4	4911.1	4914.8	4918.5	4922.2
133	4925.9	4929.6	4933.3	4937.	4940.7	4944.4	4948.1	4951.9	4955.6	4959.3
134	4963.	4966.7	4970.4	4974.1	4977.8	4981.5	4985.2	4988.9	4992.6	4996.3
135	5000.	5003.7	5007.4	5011.1	5014.8	5018.5	5022.2	5025.9	5029.6	5033.3
136	5037.	5040.7	5044.4	5048.1	5051.9	5055.6	5059.3	5063.	5066.7	5070.4
137	5074.1	5077.8	5081.5	5085.2	5088.9	5092.6	5096.3	5100.	5103.7	5107.4
138	5111.1	5114.8	5118.5	5122.2	5125.9	5129.6	5133.3	5137.	5140.7	5144.4
139	5148.1	5151.9	5155.6	5159.3	5163.	5166.7	5170.4	5174.1	5177.8	5181.5
140	5185.2	5188.9	5192.6	5196.3	5200.	5203.7	5207.4	5211.1	5214.8	5218.5
141	5222.2	5225.9	5229.6	5233.3	5237.	5240.7	5244.4	5248.1	5251.9	5255.6
142	5259.3	5263.	5266.7	5270.4	5274.1	5277.8	5281.5	5285.2	5288.9	5292.6
143	5296.3	5300.	5303.7	5307.4	5311.1	5314.8	5318.5	5322.2	5325.9	5329.6
144	5333.3	5337.	5340.7	5344.4	5348.1	5351.9	5355.6	5359.3	5363.	5366.7
145	5370.4	5374.1	5377.8	5381.5	5385.2	5388.9	5392.6	5396.3	5400.	5403.7
146	5407.4	5411.1	5414.8	5418.5	5422.2	5425.9	5429.6	5433.3	5437.	5440.7
147	5444.4	5448.1	5451.9	5455.6	5459.3	5463.	5466.7	5470.4	5474.1	5477.8
148	5481.5	5485.2	5488.9	5492.6	5496.3	5500.	5503.7	5507.4	5511.1	5514.8
149	5518.5	5522.2	5525.9	5529.6	5533.3	5537.	5540.7	5544.4	5548.1	5551.9
150	5555.6	5559.3	5563.	5566.7	5570.4	5574.1	5577.8	5581.5	5585.2	5588.9
	0.	1.	2.	3.	4.	5.	6.	7.	8.	9.

TABLE No. 4—CONTINUED.

Areas	0.1	0.2	0.25	0 3	0.4	0.5	0.6	0.7	0.75	0.8	0.9
Contents	0.4	0.7	0.9	1.1	1 5	1.9	2 2	2.6	2.8	3 0	3 3

Areas: Tens in left Column and Units at top.   Contents for 100 feet in cubic yards in body of Table.

Feet	0.	1.	2.	3.	4.	5.	6.	7.	8.	9.
151	5592.6	5596.3	5600.	5603.7	5607.4	5611.1	5614.8	5618.5	5622.2	5625.9
152	5629.6	5633.3	5637.	5640.7	5644.4	5648.1	5651.9	5655.6	5659.3	5663.
153	5666.7	5670.4	5674.1	5677.8	5681.5	5685.2	5688.9	5692.6	5696.3	5700.
154	5703.7	5707.4	5711.1	5714.8	5718.5	5722.2	5725.9	5729.6	5733.3	5737.
155	5740.7	5744.4	5748.1	5751.9	5755.6	5759.3	5763.	5766.7	5770.4	5774.1
156	5777.8	5781.5	5785.2	5788.9	5792.6	5796.3	5800.	5803.7	5807.4	5811.1
157	5814.8	5818.5	5822.2	5825.9	5829.6	5833.3	5837.	5840.7	5844.4	5848.1
158	5851.9	5855.6	5859.3	5863.	5866.7	5870.4	5874.1	5877.8	5881.5	5885.2
159	5888.9	5892.6	5896.3	5900.	5903.7	5907.4	5911.1	5914.8	5918.5	5922.2
160	5925.9	5929.6	5933.3	5937.	5940.7	5944.4	5948.1	5951.9	5955.6	5959.3
161	5963.	5966.7	5970.4	5974.1	5977.8	5981.5	5985.2	5988.9	5992.6	5996.3
162	6000.	6003.7	6007.4	6011.1	6014.8	6018.5	6022.2	6025.9	6029.6	6033.3
163	6037.	6040.7	6044.4	6048.1	6051.9	6055.6	6059.3	6063.	6066.7	6070.4
164	6074.1	6077.8	6081.5	6085.2	6088.9	6092.6	6096.3	6100.	6103.7	6107.4
165	6111.1	6114.8	6118.5	6122.2	6125.9	6129.6	6133.3	6137.	6140.7	6144.4
166	6148.1	6151.9	6155.6	6159.3	6163.	6166.7	6170.4	6174.1	6177.8	6181.5
167	6185.2	6188.9	6192.6	6196.3	6200.	6203.7	6207.4	6211.1	6214.8	6218.5
168	6222.2	6225.9	6229.6	6233.3	6237.	6240.7	6244.4	6248.1	6251.9	6255.6
169	6259.3	6263.	6266.7	6270.4	6274.1	6277.8	6281.5	6285.2	6288.9	6292.6
170	6296.3	6300.	6303.7	6307.4	6311.1	6314.8	6318.5	6322.2	6325.9	6329.6
171	6333.3	6337.	6340.7	6344.4	6348.1	6351.9	6355.6	6359.3	6363.	6366.7
172	6370.4	6374.1	6377.8	6381.5	6385.2	6388.9	6392.6	6396.3	6400.	6403.7
173	6407.4	6411.1	6414.8	6418.5	6422.2	6425.9	6429.6	6433.3	6437.	6440.7
174	6444.4	6448.1	6451.9	6455.6	6459.3	6463.	6466.7	6470.4	6474.1	6477.8
175	6481.5	6485.2	6488.9	6492.6	6496.3	6500.	6503.7	6507.4	6511.1	6514.8
176	6518.5	6522.2	6525.9	6529.6	6533.3	6537.	6540.7	6544.4	6548.1	6551.9
177	6555.6	6559.3	6563.	6566.7	6570.4	6574.1	6577.8	6581.5	6585.2	6588.9
178	6592.6	6596.3	6600.	6603.7	6607.4	6611.1	6614.8	6618.5	6622.2	6625.9
179	6629.6	6633.3	6637.	6640.7	6644.4	6648.1	6651.9	6655.6	6659.3	6663.
180	6666.7	6670.4	6674.1	6677.8	6681.5	6685.2	6688.9	6692.6	6696.3	6700.
181	6703.7	6707.4	6711.1	6714.8	6718.5	6722.2	6725.9	6729.6	6733.3	6737.
182	6740.7	6744.4	6748.1	6751.9	6755.6	6759.3	6763.	6766.7	6770.4	6774.1
183	6777.8	6781.5	6785.2	6788.9	6792.6	6796.3	6800.	6803.7	6807.4	6811.1
184	6814.8	6818.5	6822.2	6825.9	6829.6	6833.3	6837.	6840.7	6844.4	6848.1
185	6851.9	6855.6	6859.3	6863.	6866.7	6870.4	6874.1	6877.8	6881.5	6885.2
186	6888.9	6892.6	6896.3	6900.	6903.7	6907.4	6911.1	6914.8	6918.5	6922.2
187	6925.9	6929.6	6933.3	6937.	6940.7	6944.4	6948.1	6951.9	6955.6	6959.3
188	6963.	6966.7	6970.4	6974.1	6977.8	6981.5	6985.2	6988.9	6992.6	6996.3
189	7000.	7003.7	7007.4	7011.1	7014.8	7018.5	7022.2	7025.9	7029.6	7033.3
190	7037.	7040.7	7044.4	7048.1	7051.9	7055.6	7059.3	7063.	7066.7	7070.4
191	7074.1	7077.8	7081.5	7085.2	7088.9	7092.6	7096.3	7100.	7103.7	7107.4
192	7111.1	7114.8	7118.5	7122.2	7125.9	7129.6	7133.3	7137.	7140.7	7144.4
193	7148.1	7151.9	7155.6	7159.3	7163.	7166.7	7170.4	7174.1	7177.8	7181.5
194	7185.2	7188.9	7192.6	7196.3	7200.	7203.7	7207.4	7211.1	7214.8	7218.5
195	7222.2	7225.9	7229.6	7233.3	7237.	7240.7	7244.4	7248.1	7251.9	7255.6
196	7259.3	7263.	7266.7	7270.4	7274.1	7277.8	7281.5	7285.2	7288.9	7292 6
197	7296.3	7300.	7303.7	7307.4	7311.1	7314.8	7318.5	7322.2	7325 9	7329.6
198	7333.3	7337.	7340.7	7344.4	7348.1	7351.9	7355.6	7359.3	7363.	7366.7
199	7370.4	7374.1	7377.8	7381.5	7385.2	7388.9	7392.6	7396.3	7400.	7403 7
200	7407.4	7411.1	7414.8	7418.5	7422.2	7425.9	7429.6	7433.3	7437.	7440.7
	0.	1.	2.	3.	4.	5.	6.	7.	8.	9.

TABLE No. 4—CONTINUED.

Areas	0.1	0.2	0.25	0.3	0.4	0.5	0.6	0.7	0.75	0.8	0.9
Contents	0.4	0.7	0.9	1.1	1.5	1.9	2.2	2.6	2.8	3.0	3.3

Areas: Tens in left Column and Units at top. Contents for 100 feet in cubic yards in body of Table.

Feet	0.	1.	2.	3.	4.	5.	6.	7.	8.	9.
201	7444.4	7448.1	7451.9	7455.6	7459.3	7463.	7466.7	7470.4	7474.1	7477.8
202	7481.5	7485.2	7488.9	7492.6	7496.3	7500.	7503.7	7507.4	7511.1	7514.8
203	7518.5	7522.2	7525.9	7529.6	7533.3	7537.	7540.7	7544.4	7548.1	7551.9
204	7555.6	7559.3	7563.	7566.7	7570.4	7574.1	7577.8	7581.5	7585.2	7588.9
205	7592.6	7596.3	7600.	7603.7	7607.4	7611.1	7614.8	7618.5	7622.2	7625.9
206	7629.6	7633.3	7637.	7640.7	7644.4	7648.1	7651.9	7655.6	7659.3	7663.
207	7666.7	7670.4	7674.1	7677.8	7681.5	7685.2	7688.9	7692.6	7696.3	7700.
208	7703.7	7707.4	7711.1	7714.8	7718.5	7722.2	7725.9	7729.6	7733.3	7737.
209	7740.7	7744.4	7748.1	7751.9	7755.6	7759.3	7763.	7766.7	7770.4	7774.1
210	7777.8	7781.5	7785.2	7788.9	7792.6	7796.3	7800.	7803.7	7807.4	7811.1
211	7814.8	7818.5	7822.2	7825.9	7829.6	7833.3	7837.	7840.7	7844.4	7848.1
212	7851.9	7855.6	7859.3	7863.	7866.7	7870.4	7874.1	7877.8	7881.5	7885.2
213	7888.9	7892.6	7896.3	7900.	7903.7	7907.4	7911.1	7914.8	7918.5	7922.2
214	7925.9	7929.6	7933.3	7937.	7940.7	7944.4	7948.1	7951.9	7955.6	7959.3
215	7963.	7966.7	7970.4	7974.1	7977.8	7981.5	7985.2	7988.9	7992.6	7996.3
216	8000.	8003.7	8007.4	8011.1	8014.8	8018.5	8022.2	8025.9	8029.6	8033.3
217	8037.	8040.7	8044.4	8048.1	8051.9	8055.6	8059.3	8063.	8066.7	8070.4
218	8074.1	8077.8	8081.5	8085.2	8088.9	8092.6	8096.3	8100.	8103.7	8107.4
219	8111.1	8114.8	8118.5	8122.2	8125.9	8129.6	8133.3	8137.	8140.7	8144.4
220	8148.1	8151.9	8155.6	8159.3	8163.	8166.7	8170.4	8174.1	8177.8	8181.5
221	8185.2	8188.9	8192.6	8196.3	8200.	8203.7	8207.4	8211.1	8214.8	8218.5
222	8222.2	8225.9	8229.6	8233.3	8237.	8240.7	8244.4	8248.1	8251.9	8255.6
223	8259.3	8263.	8266.7	8270.4	8274.1	8277.8	8281.5	8285.2	8288.9	8292.6
224	8296.3	8300.	8303.7	8307.4	8311.1	8314.8	8318.5	8322.2	8325.9	8329.6
225	8333.3	8337.	8340.7	8344.4	8348.1	8351.9	8355.6	8359.3	8363.	8366.7
226	8370.4	8374.1	8377.8	8381.5	8385.2	8388.9	8392.6	8396.3	8400.	8403.7
227	8407.4	8411.1	8414.8	8418.5	8422.2	8425.9	8429.6	8433.3	8437.	8440.7
228	8444.4	8448.1	8451.9	8455.6	8459.3	8463.	8466.7	8470.4	8474.1	8477.8
229	8481.5	8485.2	8488.9	8492.6	8496.3	8500.	8503.7	8507.4	8511.1	8514.8
230	8518.5	8522.2	8525.9	8529.6	8533.3	8537.	8540.7	8544.4	8548.1	8551.9
231	8555.6	8559.3	8563.	8566.7	8570.4	8574.1	8577.8	8581.5	8585.2	8588.9
232	8592.6	8596.3	8600.	8603.7	8607.4	8611.1	8614.8	8618.5	8622.2	8625.9
233	8629.6	8633.3	8637.	8640.7	8644.4	8648.1	8651.9	8655.6	8659.3	8663.
234	8666.7	8670.4	8674.1	8677.8	8681.5	8685.2	8688.9	8692.6	8696.3	8700.
235	8703.7	8707.4	8711.1	8714.8	8718.5	8722.2	8725.9	8729.6	8733.3	8737.
236	8740.7	8744.4	8748.1	8751.9	8755.6	8759.3	8763.	8766.7	8770.4	8774.1
237	8777.8	8781.5	8785.2	8788.9	8792.6	8796.3	8800.	8803.7	8807.4	8811.1
238	8814.8	8818.5	8822.2	8825.9	8829.6	8833.3	8837.	8840.7	8844.4	8848.1
239	8851.9	8855.6	8859.3	8863.	8866.7	8870.4	8874.1	8877.8	8881.5	8885.2
240	8888.9	8892.6	8896.3	8900.	8903.7	8907.4	8911.1	8914.8	8918.5	8922.2
241	8925.9	8929.6	8933.3	8937.	8940.7	8944.4	8948.1	8951.9	8955.6	8959.3
242	8963.	8966.7	8970.4	8974.1	8977.8	8981.5	8985.2	8988.9	8992.6	8996.3
243	9000.	9003.7	9007.4	9011.1	9014.8	9018.5	9022.2	9025.9	9029.6	9033.3
244	9037.	9040.7	9044.4	9048.1	9051.9	9055.6	9059.3	9063.	9066.7	9070.4
245	9074.1	9077.8	9081.5	9085.2	9088.9	9092.6	9096.3	9100.	9103.7	9107.4
246	9111.1	9114.8	9118.5	9122.2	9125.9	9129.6	9133.3	9137.	9140.7	9144.4
247	9148.1	9151.9	9155.6	9159.3	9163.	9166.7	9170.4	9174.1	9177.8	9181.5
248	9185.2	9188.9	9192.6	9196.3	9200.	9203.7	9207.4	9211.1	9214.8	9218.5
249	9222.2	9225.9	9229.6	9233.3	9237.	9240.7	9244.4	9248.1	9251.9	9255.6
250	9259.3	9263.	9266.7	9270.4	9274.1	9277.8	9281.5	9285.2	9288.9	9292.6
	0.	1.	2.	3.	4.	5.	6.	7.	8.	9.

TABLE No. 4—Continued.

Areas	0.1	0.2	0.25	0.3	0.4	0.5	0.6	0.7	0.75	0.8	0.9
Contents	0.4	0.7	0.9	1.1	1.5	1.9	2.2	2.6	2.8	3.0	3.3

Areas: Tens in left Column and Units at top. Contents for 100 feet in cubic yards in body of Table.

Feet	0.	1.	2.	3.	4.	5.	6.	7.	8.	9.
251	9296.3	9300.	9303.7	9307.4	9311.1	9314.8	9318.5	9322.2	9325.9	9329.6
252	9333.3	9337.	9340.7	9344.4	9348.1	9351.9	9355.6	9359.3	9363.	9366.7
253	9370.4	9374.1	9377.8	9381.5	9385.2	9388.9	9392.6	9396.3	9400.	9403.7
254	9407.4	9411.1	9414.8	9418.5	9422.2	9425.9	9429.6	9433.3	9437.	9440.7
255	9444.4	9448.1	9451.9	9455.6	9459.3	9463.	9466.7	9470.4	9474.1	9477.8
256	9481.5	9485.2	9488.9	9492.6	9496.3	9500.	9503.7	9507.4	9511.1	9514.8
257	9518.5	9522.2	9525.9	9529.6	9533.3	9537.	9540.7	9544.4	9548.1	9551.9
258	9555.6	9559.3	9563.	9566.7	9570.4	9574.1	9577.8	9581.5	9585.2	9588.9
259	9592.6	9596.3	9600.	9603.7	9607.4	9611.1	9614.8	9618.5	9622.2	9625.9
260	9629.6	9633.3	9637.	9640.7	9644.4	9648.1	9651.9	9655.6	9659.3	9663.
261	9666.7	9670.4	9674.1	9677.8	9681.5	9685.2	9688.9	9692.6	9696.3	9700.
262	9703.7	9707.4	9711.1	9714.8	9718.5	9722.2	9725.9	9729.6	9733.3	9737.
263	9740.7	9744.4	9748.1	9751.9	9755.6	9759.3	9763.	9766.7	9770.4	9774.1
264	9777.8	9781.5	9785.2	9788.9	9792.6	9796.3	9800.	9803.7	9807.4	9811.1
265	9814.8	9818.5	9822.2	9825.9	9829.6	9833.3	9837.	9840.7	9844.4	9848.1
266	9851.9	9855.6	9859.3	9863.	9866.7	9870.4	9874.1	9877.8	9881.5	9885.2
267	9888.9	9892.6	9896.3	9900.	9903.7	9907.4	9911.1	9914.8	9918.5	9922.2
268	9925.9	9929.6	9933.3	9937.	9940.7	9944.4	9948.1	9951.9	9955.6	9959.3
269	9963.	9966.7	9970.4	9974.1	9977.8	9981.5	9985.2	9988.9	9992.6	9996.3
270	10000.	10003.7	10007.4	10011.1	10014.8	10018.5	10022.2	10025.9	10029.6	10033.3
271	10037.	10040.7	10044.4	10048.1	10051.9	10055.6	10059.3	10063.	10066.7	10070.4
272	10074.1	10077.8	10081.5	10085.2	10088.9	10092.6	10096.3	10100.	10103.7	10107.4
273	10111.1	10114.8	10118.5	10122.2	10125.9	10129.6	10133.3	10137.	10140.7	10144.4
274	10148.1	10151.9	10155.6	10159.3	10163.	10166.7	10170.4	10174.1	10177.8	10181.5
275	10185.2	10188.9	10192.6	10196.3	10200.	10203.7	10207.4	10211.1	10214.8	10218.5
276	10222.2	10225.9	10229.6	10233.3	10237.	10240.7	10244.4	10248.1	10251.9	10255.6
277	10259.3	10263.	10266.7	10270.4	10274.1	10277.8	10281.5	10285.2	10288.9	10292.6
278	10296.3	10300.	10303.7	10307.4	10311.1	10314.8	10318.5	10322.2	10325.9	10329.6
279	10333.3	10337.	10340.7	10344.4	10348.1	10351.9	10355.6	10359.3	10363.	10366.7
280	10370.4	10374.1	10377.8	10381.5	10385.2	10388.9	10392.6	10396.3	10400.	10403.7
281	10407.4	10411.1	10414.8	10418.5	10422.2	10425.9	10429.6	10433.3	10437.	10440.7
282	10444.4	10448.1	10451.9	10455.6	10459.3	10463.	10466.7	10470.4	10474.1	10477.8
283	10481.5	10485.2	10488.9	10492.6	10496.3	10500.	10503.7	10507.4	10511.1	10514.8
284	10518.5	10522.2	10525.9	10529.6	10533.3	10537.	10540.7	10544.4	10548.1	10551.9
285	10555.6	10559.3	10563.	10566.7	10570.4	10574.1	10577.8	10581.5	10585.2	10588.9
286	10592.6	10596.3	10600.	10603.7	10607.4	10611.1	10614.8	10618.5	10622.2	10625.9
287	10629.6	10633.3	10637.	10640.7	10644.4	10648.1	10651.9	10655.6	10659.3	10663.
288	10666.7	10670.4	10674.1	10677.8	10681.5	10685.2	10688.9	10692.6	10696.3	10700.
289	10703.7	10707.4	10711.1	10714.8	10718.5	10722.2	10725.9	10729.6	10733.3	10737.
290	10740.7	10744.4	10748.1	10751.9	10755.6	10759.3	10763.	10766.7	10770.4	10774.1
291	10777.8	10781.5	10785.2	10788.9	10792.6	10796.3	10800.	10803.7	10807.4	10811.1
292	10814.8	10818.5	10822.2	10825.9	10829.6	10833.3	10837.	10840.7	10844.4	10848.1
293	10851.9	10855.6	10859.3	10863.	10866.7	10870.4	10874.1	10877.8	10881.5	10885.2
294	10888.9	10892.6	10896.3	10900.	10903.7	10907.4	10911.1	10914.8	10918.5	10922.2
295	10925.9	10929.6	10933.3	10937.	10940.7	10944.4	10948.1	10951.9	10955.6	10959.3
296	10963.	10966.7	10970.4	10974.1	10977.8	10981.5	10985.2	10988.9	10992.6	10996.3
297	11000.	11003.7	11007.4	11011.1	11014.8	11018.5	11022.2	11025.9	11029.6	11033.3
298	11037.	11040.7	11044.4	11048.1	11051.9	11055.6	11059.3	11063.	11066.7	11070.4
299	11074.1	11077.8	11081.5	11085.2	11088.9	11092.6	11096.3	11100.	11103.7	11107.4
300	11111.1	11114.8	11118.5	11122.2	11125.9	11129.6	11133.3	11137.	11140.7	11144.4
	0.	1.	2.	3.	4.	5.	6.	7.	8.	9.

TABLE No. 4—CONTINUED.

Areas..........	0.1	0.2	0.25	0 3	0.4	0.5	0.6	0.7	0.75	0.8	0.9
Contents..........	0.4	0.7	0.9	1.1	1.5	1.9	2 2	2.6	2.8	3 0	3.3

Areas : Tens in left Column and Units at top.   Contents for 100 feet in cubic yards in body of Table.

Feet	0.	1.	2.	3.	4.	5.	6.	7.	8.	9.
301	11148.1	11151.9	11155.6	11159.3	11163.	11166.7	11170.4	11174.1	11177.8	11181.5
302	11185.2	11188.9	11192.6	11196.3	11200.	11203.7	11207.4	11211.1	11214.8	11218.5
303	11222.2	11225.9	11229.6	11233.3	11237.	11240.7	11244.4	11248.1	11251.9	11255.6
304	11259.3	11263.	11266.7	11270.4	11274.1	11277.8	11281.5	11285.2	11288.9	11292.6
305	11296.3	11300.	11303.7	11307.4	11311.1	11314.8	11318.5	11322.2	11325.9	11329.6
306	11333.3	11337.	11340.7	11344.4	11348.1	11351.9	11355.6	11359.3	11363.	11366.7
307	11370.4	11374.1	11377.8	11381.5	11385.2	11388.9	11392.6	11396.3	11400.	11403.7
308	11407.4	11411.1	11414.8	11418.5	11422.2	11425.9	11429.6	11433.3	11437.	11440.7
309	11444.4	11448.1	11451.9	11455.6	11459.3	11463.	11466.7	11470.4	11474.1	11477.8
310	11481.5	11485.2	11488.9	11492.6	11496.3	11500.	11503.7	11507.4	11511.1	11514.8
311	11518.5	11522.2	11525.9	11529.6	11533.3	11537.	11540.7	11544.4	11548.1	11551.9
312	11555.6	11559.3	11563.	11566.7	11570.4	11574.1	11577.8	11581.5	11585.2	11588.9
313	11592.6	11596.3	11600.	11603.7	11607.4	11611.1	11614.8	11618.5	11622.2	11625.9
314	11629.6	11633.3	11637.	11640.7	11644.4	11648.1	11651.9	11655.6	11659.3	11663.
315	11666.7	11670.4	11674.1	11677.8	11681.5	11685.2	11688.9	11692.6	11696.3	11700.
316	11703.7	11707.4	11711.1	11714.8	11718.5	11722.2	11725.9	11729.6	11733.3	11737.
317	11740.7	11744.4	11748.1	11751.9	11755.6	11759.3	11763.	11766.7	11770.4	11774.1
318	11777.8	11781.5	11785.2	11788.9	11792.6	11796.3	11800.	11803.7	11807.4	11811.1
319	11814.8	11818.5	11822.2	11825.9	11829.6	11833.3	11837.	11840.7	11844.4	11848.1
320	11851.9	11855.6	11859.3	11863.	11866.7	11870.4	11874.1	11877.8	11881.5	11885.2
321	11888.9	11892.6	11896.3	11900.	11903.7	11907.4	11911.1	11914.8	11918.5	11922.2
322	11925.9	11929.6	11933.3	11937.	11940.7	11944.4	11948.1	11951.9	11955.6	11959.3
323	11963.	11966.7	11970.4	11974.1	11977.8	11981.5	11985.2	11988.9	11992.6	11996.3
324	12000.	12003.7	12007.4	12011.1	12014.8	12018.5	12022.2	12025.9	12029.6	12033.3
325	12037.	12040.7	12044.4	12048.1	12051.9	12055.6	12059.3	12063.	12066.7	12070.4
326	12074.1	12077.8	12081.5	12085.2	12088.9	12092.6	12096.3	12100.	12103.7	12107.4
327	12111.1	12114.8	12118.5	12122.2	12125.9	12129.6	12133.3	12137.0	12140.7	12144.4
328	12148.1	12151.9	12155.6	12159.3	12163.	12166.7	12170.4	12174.1	12177.8	12181.5
329	12185.2	12188.9	12192.6	12196.3	12200.	12203.7	12207.4	12211.1	12214.8	12218.5
330	12222.2	12225.9	12229.6	12233.3	12237.	12240.7	12244.4	12248.1	12251.9	12255.6
331	12259.3	12263.	12266.7	12270.4	12274.1	12277.8	12281.5	12285.2	12288.9	12292.6
332	12296.3	12300.	12303.7	12307.4	12311.1	12314.8	12318.5	12322.2	12325.9	12329.6
333	12333.3	12337.	12340.7	12344.4	12348.1	12351.9	12355.6	12359.3	12363.	12366.7
334	12370.4	12374.1	12377.8	12381.5	12385.2	12388.9	12392.6	12396.3	12400.	12403.7
335	12407.4	12411.1	12414.8	12418.5	12422.2	12425.9	12429.6	12433.3	12437.	12440.7
336	12444.4	12448.1	12451.9	12455.6	12459.3	12463.	12466.7	12470.4	12474.1	12477.8
337	12481.5	12485.2	12488.9	12492.6	12496.3	12500.	12503.7	12507.4	12511.1	12514.8
338	12518.5	12522.2	12525.9	12529.6	12533.3	12537.	12540.7	12544.4	12548.1	12551.9
339	12555.6	12559.3	12563.	12566.7	12570.4	12574.1	12577.8	12581.5	12585.2	12588.9
340	12592.6	12596.3	12600.	12603.7	12607.4	12611.1	12614.8	12618.5	12622.2	12625.9
341	12629.6	12633.3	12637.	12640.7	12644.4	12648.1	12651.9	12655.6	12659.3	12663.
342	12666.7	12670.4	12674.1	12677.8	12681.5	12685.2	12688.9	12692.6	12696.3	12700.
343	12703.7	12707.4	12711.1	12714.8	12718.5	12722.2	12725.9	12729.6	12733.3	12737.
344	12740.7	12744.4	12748.1	12751.9	12755.6	12759.3	12763.	12766.7	12770.4	12774.1
345	12777.8	12781.5	12785.2	12788.9	12792.6	12796.3	12800.	12803.7	12807.4	12811.1
346	12814.8	12818.5	12822.2	12825.9	12829.6	12833.3	12837.	12840.7	12844.4	12848.1
347	12851.9	12855.6	12859.3	12863.	12866.7	12870.4	12874.1	12877.8	12881.5	12885.2
348	12888.9	12892.6	12896.3	12900.	12903.7	12907.4	12911.1	12914.8	12918.5	12922.2
349	12925.9	12929.6	12933.3	12937.	12940.7	12944.4	12948.1	12951.9	12955.6	12959.3
350	12963.	12966.7	12970.4	12974.1	12977.8	12981.5	12985.2	12988.9	12992.6	12996.3
	0.	1.	2.	3.	4.	5.	6.	7.	8.	9.

TABLE No. 4—CONCLUDED.

Areas............	0.1	0.2	0.25	0.3	0.4	0.5	0.6	0.7	0.75	0.8	0.9
Contents.........	0.4	0.7	0.9	1.1	1.5	1.9	2.2	2.6	2.8	3.0	3.3

Areas: Tens in left Column and Units at top. Contents for 100 feet in cubic yards in body of Table.

Feet	0.	1.	2.	3.	4.	5.	6.	7.	8.	9.
351	13000.	13003.7	13007.4	13011.1	13014.8	13018.5	13022.2	13025.9	13029.6	13033.3
352	13037.	13040.7	13044.4	13048.1	13051.9	13055.6	13059.3	13063.	13066.7	13070.4
353	13074.1	13077.8	13081.5	13085.2	13088.9	13092.6	13096.3	13100.	13103.7	13107.4
354	13111.1	13114.8	13118.5	13122.2	13125.9	13129.6	13133.3	13137.	13140.7	13144.4
355	13148.1	13151.9	13155.6	13159.3	13163.	13166.7	13170.4	13174.1	13177.8	13181.5
356	13185.2	13188.9	13192.6	13196.3	13200.	13203.7	13207.4	13211.1	13214.8	13218.5
357	13222.2	13225.9	13229.6	13233.3	13237.	13240.7	13244.4	13248.1	13251.9	13255.6
358	13259.3	13263.	13266.7	13270.4	13274.1	13277.8	13281.5	13285.2	13288.9	13292.6
359	13296.3	13300.	13303.7	13307.4	13311.1	13314.8	13318.5	13322.2	13325.9	13329.6
360	13333.3	13337.	13340.7	13344.4	13348.1	13351.9	13355.6	13359.3	13363.	13366.7
361	13370.4	13374.1	13377.8	13381.5	13385.2	13388.9	13392.6	13396.3	13400.	13403.7
362	13407.4	13411.1	13414.8	13418.5	13422.2	13425.9	13429.6	13433.3	13437.	13440.7
363	13444.4	13448.1	13451.9	13455.6	13459.3	13463.	13466.7	13470.4	13474.1	13477.8
364	13481.5	13485.2	13488.9	13492.6	13496.3	13500.	13503.7	13507.4	13511.1	13514.8
365	13518.5	13522.2	13525.9	13529.6	13533.3	13537.	13540.7	13544.4	13548.1	13551.9
366	13555.6	13559.3	13563.	13566.7	13570.4	13574.1	13577.8	13581.5	13585.2	13588.9
367	13592.6	13596.3	13600.	13603.7	13607.4	13611.1	13614.8	13618.5	13622.2	13625.9
368	13629.6	13633.3	13637.	13640.7	13644.4	13648.1	13651.9	13655.6	13659.3	13663.
369	13666.7	13670.4	13674.1	13677.8	13681.5	13685.2	13688.9	13692.6	13696.3	13700.
370	13703.7	13707.4	13711.1	13714.8	13718.5	13722.2	13725.9	13729.6	13733.3	13737.
371	13740.7	13744.4	13748.1	13751.9	13755.6	13759.3	13763.	13766.7	13770.4	13774.1
372	13777.8	13781.5	13785.2	13788.9	13792.6	13796.3	13800.	13803.7	13807.4	13811.1
373	13814.8	13818.5	13822.2	13825.9	13829.6	13833.3	13837.	13840.7	13844.4	13848.1
374	13851.9	13855.6	13859.3	13863.	13866.7	13870.4	13874.1	13877.8	13881.5	13885.2
375	13888.9	13892.6	13896.3	13900.	13903.7	13907.4	13911.1	13914.8	13918.5	13922.2
376	13925.9	13929.6	13933.3	13937.	13940.7	13944.4	13948.1	13951.9	13955.6	13959.3
377	13963.	13966.7	13970.4	13974.1	13977.8	13981.5	13985.2	13988.9	13992.6	13996.3
378	14000.	14003.7	14007.4	14011.1	14014.8	14018.5	14022.2	14025.9	14029.6	14033.3
379	14037.	14040.7	14044.4	14048.1	14051.9	14055.6	14059.3	14063.	14066.7	14070.4
380	14074.1	14077.8	14081.5	14085.2	14088.9	14092.6	14096.3	14100.	14103.7	14107.4
381	14111.1	14114.8	14118.5	14122.2	14125.9	14129.6	14133.3	14137.4	14140.7	14144.4
382	14148.1	14151.9	14155.6	14159.3	14163.	14166.7	14170.4	14174.1	14177.8	14181.5
383	14185.2	14188.9	14192.6	14196.3	14200.	14203.7	14207.4	14211.1	14214.8	14218.5
384	14222.2	14225.9	14229.6	14233.3	14237.	14240.7	14244.4	14248.1	14251.9	14255.6
385	14259.3	14263.	14266.7	14270.4	14274.1	14277.8	14281.5	14285.2	14288.9	14292.6
386	14296.3	14300.	14303.7	14307.4	14311.1	14314.8	14318.5	14322.2	14325.9	14329.6
387	14333.3	14337.	14340.7	14344.4	14348.1	14351.9	14355.6	14359.3	14363.	14366.7
388	14370.4	14374.1	14377.8	14381.5	14385.2	14388.9	14392.6	14396.3	14400.	14403.7
389	14407.4	14411.1	14414.8	14418.5	14422.2	14425.9	14429.6	14433.3	14437.	14440.7
390	14444.4	14448.1	14451.9	14455.6	14459.3	14463.	14466.7	14470.4	14474.1	14477.8
391	14481.5	14485.2	14488.9	14492.6	14496.3	14500.	14503.7	14507.4	14511.1	14514.8
392	14518.5	14522.2	14525.9	14529.6	14533.3	14537.	14540.7	14544.4	14548.1	14551.9
393	14555.6	14559.3	14563.	14566.7	14570.4	14574.1	14577.8	14581.5	14585.2	14588.9
394	14592.6	14596.3	14600.	14603.7	14607.4	14611.1	14614.8	14618.5	14622.2	14625.9
395	14629.6	14633.3	14637.	14640.7	14644.4	14648.1	14651.9	14655.6	14659.3	14663.
396	14666.7	14670.4	14674.1	14677.8	14681.5	14685.2	14688.9	14692.6	14696.3	14700.
397	14703.7	14707.4	14711.1	14714.8	14718.5	14722.2	14725.9	14729.6	14733.3	14737.
398	14740.7	14744.4	14748.1	14751.9	14755.6	14759.3	14763.	14766.7	14770.4	14774.1
399	14777.8	14781.5	14785.2	14788.9	14792.6	14796.3	14800.	14803.7	14807.4	14811.1
400	14814.8	14818.5	14822.2	14825.9	14829.6	14833.3	14837.	14840.7	14844.4	14848.1
	0.	1.	2.	3.	4.	5.	6.	7.	8.	9.

TABLE No. 5.

Minus Corrections corresponding to $N \sim N'$, or $n \sim n'$, and general for all side slopes. For computation by average Areas.

Difference of Correction numbers in feet and tenths in left column and at top ; Correction in cubic yards for 100 ft. in body of Table.

Feet	0.	1.	2.	3.	4.	5.	6.	7.	8.	9.
0	0.0	0.0	0.0	0.1	0.1	0.2	0.2	0.3	0.4	0.5
1	0.6	0.7	0.9	1.0	1.2	1.4	1.6	1.8	2.0	2.2
2	2.5	2.7	3.0	3.3	3.6	3.9	4.2	4.5	4.8	5.2
3	5.6	5.9	6.3	6.7	7.1	7.6	8.0	8.5	8.9	9.4
4	9.9	10.4	10.9	11.4	12.0	12.5	13.1	13.6	14.2	14.8
5	15.4	16.1	16.7	17.3	18.0	18.7	19.4	20.1	20.8	21.5
6	22.2	23.0	23.7	24.5	25.3	26.1	26.9	27.7	28.5	29.4
7	30.2	31.1	32.0	32.9	33.8	34.7	35.7	36.6	37.6	38.5
8	39.5	40.5	41.5	42.5	43.6	44.6	45.7	46.7	47.8	48.9
9	50.0	51.1	52.2	53.4	54.5	55.7	56.9	58.1	59.3	60.5
10	61.7	63.0	64.2	65.5	66.8	68.1	69.4	70.7	72.0	73.3
11	74.7	76.1	77.4	78.8	80.2	81.6	83.1	84.5	86.0	87.4
12	88.9	90.4	91.9	93.4	94.9	96.5	98.0	99.6	101.1	102.7
13	104.3	105.9	107.6	109.2	110.8	112.5	114.2	115.9	117.6	119.3
14	121.0	122.7	124.5	126.2	128.0	129.8	131.6	133.4	135.2	137.0
15	138.9	140.7	142.6	144.5	146.4	148.3	150.2	152.2	154.1	156.1
16	158.0	160.0	162.0	164.0	166.0	168.1	170.1	172.2	174.2	176.3
17	178.4	180.5	182.6	184.7	186.9	189.0	191.2	193.4	195.6	197.8
18	200.0	202.2	204.5	206.7	209.0	211.3	213.6	215.9	218.2	220.5
19	222.8	225.2	227.6	229.9	232.3	234.7	237.1	239.6	242.0	244.5
20	246.9	249.4	251.9	254.4	256.9	259.4	262.0	264.5	267.1	269.6
21	272.2	274.8	277.4	280.1	282.7	285.3	288.0	290.7	293.4	296.1
22	298.8	301.5	304.2	307.0	309.7	312.5	315.3	318.1	320.9	323.7
23	326.5	329.4	332.2	335.1	338.0	340.9	343.8	346.7	349.7	352.6
24	355.6	358.5	361.5	364.5	367.5	370.5	373.6	376.6	379.7	382.7
25	385.8	388.9	392.0	395.1	398.2	401.4	404.5	407.7	410.9	414.1
26	417.3	420.5	423.7	427.0	430.2	433.5	436.8	440.1	443.4	446.7
27	450.0	453.3	456.7	460.1	463.4	466.8	470.2	473.6	477.1	480.5
28	484.0	487.4	490.9	494.4	497.9	501.4	504.9	508.5	512.0	515.6
29	519.1	522.7	526.3	529.9	533.6	537.2	540.8	544.5	548.2	551.9
30	555.6	559.3	563.0	566.7	570.5	574.2	578.0	581.8	585.6	589.4
31	593.2	597.0	600.9	604.7	608.6	612.5	616.4	620.3	624.2	628.2
32	632.1	636.1	640.0	644.0	648.0	652.0	656.0	660.1	664.1	668.2
33	672.2	676.3	680.4	684.5	688.6	692.7	696.9	701.0	705.2	709.4
34	713.6	717.8	722.0	726.2	730.5	734.7	739.0	743.3	747.6	751.9
35	756.2	760.5	764.8	769.2	773.6	777.9	782.3	786.7	791.1	795.6
36	800.0	804.5	808.9	813.4	817.9	822.4	826.9	831.4	836.0	840.5
37	845.1	849.6	854.2	858.8	863.4	868.1	872.7	877.3	882.0	886.7
38	891.4	896.1	900.8	905.5	910.2	915.0	919.7	924.5	929.3	934.1
39	938.9	943.7	948.5	953.4	958.2	963.1	968.0	972.9	977.8	982.7
40	987.7	992.6	997.6	1002.5	1007.5	1012.5	1017.5	1022.5	1027.6	1032.6
41	1037.7	1042.7	1047.8	1052.9	1058.0	1063.1	1068.2	1073.4	1078.5	1083.7
42	1088.9	1094.1	1099.3	1104.5	1109.7	1115.0	1120.2	1125.5	1130.8	1136.1
43	1141.4	1146.7	1152.0	1157.3	1162.7	1168.1	1173.4	1178.8	1184.2	1189.6
44	1195.1	1200.5	1206.0	1211.4	1216.9	1222.4	1227.9	1233.4	1238.9	1244.5
45	1250.0	1255.6	1261.1	1266.7	1272.3	1277.9	1283.6	1289.2	1294.8	1300.5
46	1306.2	1311.9	1317.6	1323.3	1329.0	1334.7	1340.5	1346.2	1352.0	1357.8
47	1363.6	1369.4	1375.2	1381.0	1386.9	1392.7	1398.6	1404.5	1410.4	1416.3
48	1422.2	1428.2	1434.1	1440.1	1446.0	1452.0	1458.0	1464.0	1470.0	1476.1
49	1482.1	1488.2	1494.2	1500.3	1506.4	1512.5	1518.6	1524.7	1530.9	1537.0
50	1543.2	1549.4	1555.6	1561.8	1568.0	1574.2	1580.5	1586.7	1593.0	1599.3
	0.	1.	2.	3.	4.	5.	6.	7.	8.	9.

TABLE No. 5—CONCLUDED.

Minus Corrections corresponding to $N \sim N'$, or $n \sim n'$, and general for all side slopes. For computation by average Areas.

Difference of Correction numbers in feet and tenths in left column and at top; Correction in cubic yards for 100 ft. in body of Table.

Feet	.0	.1	.2	.3	.4	.5	.6	.7	.8	.9
51	1605.6	1611.9	1618.2	1624.5	1630.8	1637.2	1643.6	1649.9	1656.3	1662.7
52	1669.1	1675.6	1682.0	1688.5	1694.9	1701.4	1707.9	1714.4	1720.9	1727.4
53	1734.0	1740.5	1747.1	1753.6	1760.2	1766.8	1773.4	1780.1	1786.7	1793.3
54	1800.0	1806.7	1813.4	1820.1	1826.8	1833.5	1840.2	1847.0	1853.7	1860.5
55	1867.3	1874.1	1880.9	1887.7	1894.5	1901.4	1908.2	1915.1	1922.0	1928.9
56	1935.8	1942.7	1949.7	1956.6	1963.6	1970.5	1977.5	1984.5	1991.5	1998.5
57	2005.6	2012.6	2019.7	2026.7	2033.8	2040.9	2048.0	2055.1	2062.2	2069.4
58	2076.5	2083.7	2090.9	2098.1	2105.3	2112.5	2119.7	2127.0	2134.2	2141.5
59	2148.8	2156.1	2163.4	2170.7	2178.0	2185.3	2192.7	2200.1	2207.4	2214.8
60	2222.2	2229.6	2237.1	2244.5	2252.0	2259.4	2266.9	2274.4	2281.9	2289.4
61	2296.9	2304.5	2312.0	2319.6	2327.1	2334.7	2342.3	2349.9	2357.6	2365.2
62	2372.8	2380.5	2388.2	2395.9	2403.6	2411.3	2419.0	2426.7	2434.5	2442.2
63	2450.0	2457.8	2465.6	2473.4	2481.2	2489.0	2496.9	2504.7	2512.6	2520.5
64	2528.4	2536.3	2544.2	2552.2	2560.1	2568.1	2576.0	2584.0	2592.0	2600.0
65	2608.0	2615.1	2624.1	2632.2	2640.2	2648.3	2656.4	2664.5	2672.6	2680.7
66	2688.9	2697.0	2705.2	2713.4	2721.6	2729.8	2738.0	2746.2	2754.5	2762.7
67	2771.0	2779.3	2787.6	2795.9	2804.2	2812.5	2820.8	2829.2	2837.6	2845.9
68	2854.3	2862.7	2871.1	2879.6	2888.0	2896.5	2904.9	2913.4	2921.9	2930.4
69	2938.9	2947.4	2956.0	2964.5	2973.1	2981.6	2990.2	2998.8	3007.4	3016.1
70	3024.7	3033.3	3042.0	3050.7	3059.4	3068.1	3076.8	3085.5	3094.2	3103.0
71	3111.7	3120.5	3129.3	3138.1	3146.9	3155.7	3164.5	3173.4	3182.2	3191.1
72	3200.0	3208.9	3217.8	3226.7	3235.7	3244.6	3253.6	3262.5	3271.5	3280.5
73	3289.5	3298.5	3307.6	3316.6	3325.7	3334.7	3343.8	3352.9	3362.0	3371.1
74	3380.2	3389.4	3398.5	3407.7	3416.9	3426.1	3435.3	3444.5	3453.7	3463.0
75	3472.2	3481.5	3490.8	3500.1	3509.4	3518.7	3528.0	3537.3	3546.7	3556.1
76	3565.4	3574.8	3584.2	3593.6	3603.1	3612.5	3622.0	3631.4	3640.9	3650.4
77	3659.9	3669.4	3678.9	3688.5	3698.0	3707.6	3717.1	3726.7	3736.3	3745.9
78	3755.6	3765.2	3774.8	3784.5	3794.2	3803.9	3813.6	3823.3	3833.0	3842.7
79	3852.5	3862.2	3872.0	3881.8	3891.6	3901.4	3911.2	3921.0	3930.9	3940.7
80	3950.6	3960.5	3970.4	3980.3	3990.2	4000.2	4010.1	4020.1	4030.0	4040.0
81	4050.0	4060.0	4070.0	4080.1	4090.1	4100.2	4110.2	4120.3	4130.4	4140.5
82	4150.6	4160.7	4170.9	4181.0	4191.2	4201.4	4211.6	4221.8	4232.0	4242.2
83	4252.5	4262.7	4273.0	4283.3	4293.6	4303.9	4314.2	4324.5	4334.8	4345.2
84	4355.6	4365.9	4376.3	4386.7	4397.1	4407.6	4418.0	4428.5	4438.9	4449.4
85	4459.9	4470.4	4480.9	4491.4	4502.0	4512.5	4523.1	4533.6	4544.2	4554.8
86	4565.4	4576.1	4586.7	4597.3	4608.0	4618.7	4629.4	4640.1	4650.8	4661.5
87	4672.2	4683.0	4693.7	4704.5	4715.3	4726.1	4736.9	4747.7	4758.5	4769.4
88	4780.2	4791.1	4802.0	4812.9	4823.8	4834.7	4845.7	4856.6	4867.6	4878.5
89	4889.5	4900.5	4911.5	4922.5	4933.6	4944.6	4955.7	4966.7	4977.8	4988.9
90	5000.0	5011.1	5022.2	5033.4	5044.5	5055.7	5066.9	5078.1	5089.3	5100.5
91	5111.7	5123.0	5134.2	5145.5	5156.8	5168.1	5179.4	5190.7	5202.0	5213.3
92	5224.7	5236.1	5247.4	5258.8	5270.2	5281.6	5293.1	5304.5	5316.0	5327.4
93	5338.9	5350.4	5361.9	5373.4	5384.9	5396.5	5408.0	5419.6	5431.1	5442.7
94	5454.3	5465.9	5477.6	5489.2	5500.8	5512.5	5524.2	5535.9	5547.6	5559.3
95	5571.0	5582.7	5594.5	5606.2	5618.0	5629.8	5641.6	5653.4	5665.2	5677.0
96	5688.9	5700.7	5712.6	5724.5	5736.4	5748.3	5760.2	5772.2	5784.1	5796.1
97	5808.0	5820.0	5832.0	5844.0	5856.0	5868.1	5880.1	5892.2	5904.2	5916.3
98	5928.4	5940.5	5952.6	5964.7	5976.9	5989.0	6001.2	6013.4	6025.6	6037.8
99	6050.0	6062.2	6074.5	6086.7	6099.0	6111.3	6123.6	6135.9	6148.2	6160.5
100	6172.8	6185.2	6197.6	6209.9	6222.3	6234.7	6247.1	6259.6	6272.0	6284.5
	.0	.1	.2	.3	.4	.5	.6	.7	.8	.9

TABLE No. 6.—LEVEL CUTTINGS. $\frac{s+s'}{2} = \frac{1}{5}$; $b = 16$ feet.

道	.0	.1	.2	.3	.4	.5	.6	.7	.8	.9
0	0.0	5.9	11.9	17.8	23.8	29.8	35.8	41.8	47.9	53.9
1	60.0	66.1	72.2	78.3	84.4	90.6	96.7	102.9	109.1	115.3
2	121.5	127.7	134.0	140.2	146.5	152.8	159.1	165.4	171.7	178.1
3	184.4	190.8	197.2	203.6	210.0	216.5	222.9	229.4	235.9	242.4
4	248.9	255.4	262.0	268.5	275.1	281.7	288.3	294.9	301.5	308.2
5	314.8	321.5	328.2	334.9	341.6	348.3	355.1	361.8	368.6	375.4
6	382.2	389.0	395.9	402.7	409.6	416.5	423.4	430.3	437.2	444.2
7	451.1	458.1	465.1	472.1	479.1	486.1	493.2	500.2	507.3	514.4
8	521.5	528.6	535.7	542.9	550.0	557.2	564.4	571.6	578.8	586.1
9	593.3	600.6	607.9	615.2	622.5	629.8	637.2	644.5	651.9	659.3
10	666.7	674.1	681.5	689.0	696.4	703.9	711.4	718.9	726.4	733.9
11	741.5	749.0	756.6	764.2	771.8	779.4	787.1	794.7	802.4	810.1
12	817.8	825.5	833.2	841.0	848.7	856.5	864.3	872.1	879.9	887.7
13	895.6	903.4	911.3	919.2	927.1	935.0	942.9	950.9	958.8	966.8
14	974.8	982.8	990.8	998.9	1007	1015	1023	1031	1039	1047
15	1056	1064	1072	1080	1088	1096	1105	1113	1121	1129
16	1138	1146	1154	1163	1171	1179	1188	1196	1205	1213
17	1221	1230	1238	1247	1255	1264	1272	1281	1290	1298
18	1307	1315	1324	1333	1341	1350	1358	1367	1376	1385
19	1393	1402	1411	1420	1428	1437	1446	1455	1464	1473
20	1482	1490	1499	1508	1517	1526	1535	1544	1553	1562
21	1571	1580	1589	1598	1607	1616	1626	1635	1644	1653
22	1662	1671	1681	1690	1699	1708	1718	1727	1736	1745
23	1755	1764	1774	1783	1792	1802	1811	1821	1830	1839
24	1849	1858	1868	1877	1887	1896	1906	1916	1925	1935
25	1944	1954	1964	1973	1983	1993	2002	2012	2022	2032
26	2041	2051	2061	2071	2081	2091	2100	2110	2120	2130
27	2140	2150	2160	2170	2180	2190	2200	2210	2220	2230
28	2240	2250	2260	2270	2280	2291	2301	2311	2321	2331
29	2341	2352	2362	2372	2382	2393	2403	2413	2424	2434
30	2444	2455	2465	2476	2486	2496	2507	2517	2528	2538
31	2549	2559	2570	2581	2591	2602	2612	2623	2634	2644
32	2655	2665	2676	2687	2698	2708	2719	2730	2741	2751
33	2762	2773	2784	2795	2806	2816	2827	2838	2849	2860
34	2871	2882	2893	2904	2915	2926	2937	2948	2959	2970
35	2981	2993	3004	3015	3026	3037	3048	3060	3071	3082
36	3093	3105	3116	3127	3138	3150	3161	3173	3184	3195
37	3207	3218	3230	3241	3252	3264	3275	3287	3298	3310
38	3321	3333	3345	3356	3368	3379	3391	3403	3414	3426
39	3438	3449	3461	3473	3485	3496	3508	3520	3532	3544
40	3556	3567	3579	3581	3593	3605	3617	3629	3641	3653
41	3675	3687	3699	3711	3723	3735	3747	3759	3771	3783
42	3796	3808	3820	3832	3844	3856	3869	3881	3893	3905
43	3918	3930	3942	3955	3967	3979	3992	4004	4017	4029
44	4041	4054	4066	4079	4091	4104	4116	4129	4142	4154
45	4167	4179	4192	4205	4217	4230	4242	4255	4268	4281
46	4293	4306	4319	4332	4344	4357	4370	4383	4396	4409
47	4421	4434	4447	4460	4473	4486	4499	4512	4525	4538
48	4551	4564	4577	4590	4603	4616	4630	4643	4656	4669
49	4682	4695	4709	4722	4735	4748	4762	4775	4788	4801
50	4815	4828	4842	4855	4868	4882	4895	4909	4922	4935
51	4949	4962	4976	4989	5003	5016	5030	5044	5057	5071
52	5084	5098	5112	5125	5139	5153	5166	5180	5194	5208
53	5221	5235	5249	5263	5277	5291	5304	5318	5332	5346
54	5360	5374	5388	5402	5416	5430	5444	5458	5472	5486
55	5500	5514	5528	5542	5556	5571	5585	5599	5613	5627
56	5641	5656	5670	5684	5698	5713	5727	5741	5756	5770
57	5784	5799	5813	5828	5842	5856	5871	5885	5900	5914
58	5929	5943	5958	5973	5987	6002	6016	6031	6046	6060
59	6075	6089	6104	6119	6134	6148	6163	6178	6193	6207
60	6222	6237	6252	6267	6282	6296	6311	6326	6341	6356
	.0	.1	.2	.3	.4	.5	.6	.7	.8	.9

TABLE No. 7.—LEVEL CUTTINGS. $\frac{s+s'}{2}=\frac{1}{5}$; $b=28$ feet.

F.	.0	.1	.2	.3	.4	.5	.6	.7	.8	.9
0	0.0	10.4	20.8	31.2	41.6	52.0	62.5	73.0	83.4	93.9
1	104.4	115.0	125.5	136.1	146.6	157.2	167.8	178.4	189.1	199.7
2	210.4	221.0	231.7	242.4	253.2	263.9	274.6	285.4	296.2	307.0
3	317.8	328.6	339.4	350.3	361.2	372.0	382.9	393.8	404.8	415.7
4	426.7	437.6	448.6	459.6	470.6	481.7	492.7	503.8	514.8	525.9
5	537.0	548.2	559.3	570.4	581.6	592.8	604.0	615.2	626.4	637.6
6	648.9	660.2	671.4	682.7	694.0	705.4	716.7	728.1	739.4	750.8
7	762.2	773.6	785.1	796.5	808.0	819.4	830.9	842.4	854.0	865.5
8	877.0	888.6	900.2	911.8	923.4	935.0	946.6	958.3	970.0	981.6
9	993.3	1005	1017	1029	1040	1052	1064	1076	1087	1099
10	1111	1123	1135	1147	1159	1171	1182	1194	1206	1218
11	1230	1242	1254	1266	1278	1291	1303	1315	1327	1339
12	1351	1363	1375	1388	1400	1412	1424	1437	1449	1461
13	1473	1486	1498	1510	1523	1535	1547	1560	1572	1585
14	1597	1609	1622	1634	1647	1659	1672	1685	1697	1710
15	1722	1735	1747	1760	1773	1785	1798	1811	1823	1836
16	1849	1862	1874	1887	1900	1913	1926	1938	1951	1964
17	1977	1990	2003	2016	2029	2042	2055	2068	2081	2094
18	2107	2120	2133	2146	2159	2172	2185	2198	2211	2225
19	2238	2251	2264	2277	2291	2304	2317	2330	2344	2357
20	2370	2384	2397	2410	2424	2437	2451	2464	2478	2491
21	2504	2518	2531	2545	2558	2572	2586	2599	2613	2626
22	2640	2654	2667	2681	2695	2708	2722	2736	2750	2763
23	2777	2791	2805	2818	2832	2846	2860	2874	2888	2902
24	2916	2929	2943	2957	2971	2985	2999	3013	3027	3041
25	3056	3070	3084	3098	3112	3126	3140	3154	3169	3183
26	3197	3211	3226	3240	3254	3268	3283	3297	3311	3326
27	3340	3354	3369	3383	3398	3412	3426	3441	3455	3470
28	3484	3499	3514	3528	3543	3557	3572	3586	3601	3616
29	3630	3645	3660	3674	3689	3704	3719	3733	3748	3763
30	3778	3793	3807	3822	3837	3852	3867	3882	3897	3912
31	3927	3942	3957	3972	3987	4002	4017	4032	4047	4062
32	4077	4092	4107	4122	4138	4153	4168	4183	4198	4214
33	4229	4244	4259	4275	4290	4305	4321	4336	4351	4367
34	4382	4398	4413	4429	4444	4459	4475	4490	4506	4521
35	4537	4553	4568	4584	4599	4615	4631	4646	4662	4678
36	4693	4709	4725	4741	4756	4772	4788	4804	4819	4835
37	4851	4867	4883	4899	4915	4931	4946	4962	4978	4994
38	5010	5026	5042	5058	5074	5091	5107	5123	5139	5155
39	5171	5187	5203	5220	5236	5252	5268	5285	5301	5317
40	5333	5350	5366	5382	5399	5415	5431	5448	5464	5481
41	5497	5513	5530	5546	5563	5579	5596	5613	5629	5646
42	5662	5679	5695	5712	5729	5745	5762	5779	5795	5812
43	5829	5846	5862	5879	5896	5913	5930	5946	5963	5980
44	5997	6014	6031	6048	6065	6082	6099	6116	6133	6150
45	6167	6184	6201	6218	6235	6252	6269	6286	6303	6321
46	6338	6355	6372	6389	6407	6424	6441	6458	6476	6493
47	6510	6528	6545	6562	6580	6597	6615	6632	6650	6667
48	6684	6702	6719	6737	6754	6772	6790	6807	6825	6842
49	6860	6878	6895	6913	6931	6948	6966	6984	7002	7019
50	7037	7055	7073	7090	7108	7126	7144	7162	7180	7198
51	7216	7233	7251	7269	7287	7305	7323	7341	7359	7377
52	7396	7414	7432	7450	7468	7486	7504	7522	7541	7559
53	7577	7595	7614	7632	7650	7668	7687	7705	7723	7742
54	7760	7778	7797	7815	7834	7852	7870	7889	7907	7926
55	7944	7963	7982	8000	8019	8037	8056	8074	8093	8112
56	8130	8149	8168	8186	8205	8224	8243	8261	8280	8299
57	8318	8337	8355	8374	8393	8412	8431	8450	8469	8488
58	8507	8526	8545	8564	8583	8602	8621	8640	8659	8678
59	8697	8716	8735	8754	8774	8793	8812	8831	8850	8870
60	8889	8908	8927	8947	8966	8985	9005	9024	9043	9063
	.0	.1	.2	.3	.4	.5	.6	.7	.8	.9

TABLE No. 8.

Plus Corrections for $\frac{s+s'}{2}=\frac{1}{5}$.

Feet.	0.	1.	2.	3.	4.	5.	6.	7.	8.	9.
0	0.0	0.0	0.0	0.0	0.0	0.0	0.0	0.0	0.0	0.1
1	0.1	0.1	0.1	0.1	0.1	0.1	0.2	0.2	0.2	0.2
2	0.3	0.3	0.3	0.3	0.4	0.4	0.4	0.5	0.5	0.5
3	0.6	0.6	0.6	0.7	0.7	0.8	0.8	0.9	0.9	0.9
4	1.0	1.0	1.1	1.1	1.2	1.3	1.3	1.4	1.4	1.5
5	1.5	1.6	1.7	1.7	1.8	1.9	1.9	2.0	2.1	2.2
6	2.2	2.3	2.4	2.5	2.5	2.6	2.7	2.8	2.9	2.9
7	3.0	3.1	3.2	3.3	3.4	3.5	3.6	3.7	3.8	3.9
8	4.0	4.1	4.2	4.3	4.4	4.5	4.6	4.7	4.8	4.9
9	5.0	5.1	5.2	5.3	5.5	5.6	5.7	5.8	5.9	6.1
10	6.2	6.3	6.4	6.6	6.7	6.8	6.9	7.1	7.2	7.3
11	7.5	7.6	7.7	7.9	8.0	8.2	8.3	8.5	8.6	8.7
12	8.9	9.0	9.2	9.3	9.5	9.7	9.8	10.0	10.1	10.3
13	10.4	10.6	10.8	10.9	11.1	11.3	11.4	11.6	11.8	11.9
14	12.1	12.3	12.5	12.6	12.8	13.0	13.2	13.3	13.5	13.7
15	13.9	14.1	14.3	14.5	14.6	14.8	15.0	15.2	15.4	15.6
16	15.8	16.0	16.2	16.4	16.6	16.8	17.0	17.2	17.4	17.6
17	17.8	18.1	18.3	18.5	18.7	18.9	19.1	19.3	19.6	19.8
18	20.0	20.2	20.5	20.7	20.9	21.1	21.4	21.6	21.8	22.1
19	22.3	22.5	22.8	23.0	23.2	23.5	23.7	24.0	24.2	24.5
20	24.7	24.9	25.2	25.4	25.7	25.9	26.2	26.5	26.7	27.0
21	27.2	27.5	27.7	28.0	28.3	28.5	28.8	29.1	29.3	29.6
22	29.9	30.2	30.4	30.7	31.0	31.3	31.5	31.8	32.1	32.4
23	32.7	32.9	33.2	33.5	33.8	34.1	34.4	34.7	35.0	35.3
24	35.6	35.9	36.2	36.5	36.8	37.1	37.4	37.7	38.0	38.3
25	38.6	38.9	39.2	39.5	39.8	40.1	40.5	40.8	41.1	41.4
26	41.7	42.1	42.4	42.7	43.0	43.4	43.7	44.0	44.3	44.7
27	45.0	45.3	45.7	46.0	46.3	46.7	47.0	47.4	47.7	48.1
28	48.4	48.7	49.1	49.4	49.8	50.1	50.5	50.9	51.2	51.6
29	51.9	52.3	52.6	53.0	53.4	53.7	54.1	54.5	54.8	55.2
30	55.6	55.9	56.3	56.7	57.1	57.4	57.8	58.2	58.6	58.9
31	59.3	59.7	60.1	60.5	60.9	61.3	61.6	62.0	62.4	62.8
32	63.2	63.6	64.0	64.4	64.8	65.2	65.6	66.0	66.4	66.8
33	67.2	67.6	68.0	68.5	68.9	69.3	69.7	70.1	70.5	70.9
34	71.4	71.8	72.2	72.6	73.1	73.5	73.9	74.3	74.8	75.2
35	75.6	76.1	76.5	76.9	77.4	77.8	78.2	78.7	79.1	79.6
36	80.0	80.5	80.9	81.3	81.8	82.2	82.7	83.1	83.6	84.1
37	84.5	85.0	85.4	85.9	86.3	86.8	87.3	87.7	88.2	88.7
38	89.1	89.6	90.1	90.6	91.0	91.5	92.0	92.5	92.9	93.4
39	93.9	94.4	94.9	95.3	95.8	96.3	96.8	97.3	97.8	98.3
40	98.8	99.3	99.8	100.3	100.8	101.3	101.8	102.3	102.8	103.3
	0.	1.	2.	3.	4.	5.	6.	7.	8.	9.

NOTE.—The quantities in the above table multiplied by 2 give the minus corrections for $\frac{s+s'}{2} = \frac{1}{5}$.

TABLE No. 9.—LEVEL CUTTINGS. $\frac{s+s'}{2}=\frac{1}{2}$; $b = 16$ feet.

Ft.	0.	1.	2.	3.	4.	5.	6.	7.	8.	9.
0	0.0	5.9	11.9	17.9	24.0	30.1	36.2	42.4	48.6	54.8
1	61.1	67.4	73.8	80.2	86.6	93.1	99.6	106.1	112.7	119.3
2	125.9	132.6	139.3	146.1	152.9	159.7	166.6	173.5	180.4	187.4
3	194.4	201.5	208.6	215.7	222.9	230.1	237.3	244.6	251.9	259.3
4	266.7	274.1	281.6	289.1	296.6	304.2	311.8	319.4	327.1	334.8
5	342.6	350.4	358.2	366.1	374.0	381.9	389.9	397.9	406.0	414.1
6	422.2	430.4	438.6	446.8	455.1	463.4	471.8	480.2	488.6	497.1
7	505.6	514.1	522.7	531.3	539.9	548.6	557.3	566.1	574.9	583.7
8	592.6	601.5	610.4	619.4	628.4	637.5	646.6	655.7	664.9	674.1
9	683.3	692.6	701.9	711.3	720.7	730.1	739.6	749.1	758.6	768.2
10	777.8	787.4	797.1	806.8	816.6	826.4	836.2	846.1	856.0	865.9
11	875.9	885.9	896.0	906.1	916.2	926.4	936.6	946.8	957.1	967.4
12	977.8	988.2	998.6	1009	1020	1030	1041	1051	1062	1073
13	1083	1094	1105	1116	1127	1138	1148	1159	1170	1182
14	1193	1204	1215	1226	1237	1249	1260	1271	1283	1294
15	1306	1317	1329	1340	1352	1363	1375	1387	1399	1410
16	1422	1434	1446	1458	1470	1482	1494	1506	1518	1530
17	1543	1555	1567	1579	1592	1604	1617	1629	1642	1654
18	1667	1679	1692	1705	1717	1730	1743	1756	1769	1782
19	1794	1807	1820	1834	1847	1860	1873	1886	1899	1913
20	1926	1939	1953	1966	1980	1993	2007	2020	2034	2047
21	2061	2075	2089	2102	2116	2130	2144	2158	2172	2186
22	2200	2214	2228	2242	2257	2271	2285	2299	2314	2328
23	2343	2357	2372	2386	2401	2415	2430	2445	2459	2474
24	2489	2504	2519	2534	2548	2563	2578	2594	2609	2624
25	2639	2654	2669	2685	2700	2715	2731	2746	2762	2777
26	2793	2808	2824	2839	2855	2871	2887	2902	2918	2934
27	2950	2966	2982	2998	3014	3030	3046	3062	3079	3095
28	3111	3127	3144	3160	3177	3193	3210	3226	3243	3259
29	3276	3293	3309	3326	3343	3360	3377	3394	3410	3427
30	3444	3462	3479	3496	3513	3530	3547	3565	3582	3599
31	3617	3634	3652	3669	3687	3704	3722	3739	3757	3775
32	3793	3810	3828	3846	3864	3882	3900	3918	3936	3954
33	3972	3990	4009	4027	4045	4063	4082	4100	4119	4137
34	4156	4174	4193	4211	4230	4249	4267	4286	4305	4324
35	4343	4362	4380	4399	4418	4438	4457	4476	4495	4514
36	4533	4553	4572	4591	4611	4630	4650	4669	4689	4708
37	4728	4747	4767	4787	4807	4826	4846	4866	4886	4906
38	4926	4946	4966	4986	5006	5026	5047	5067	5087	5107
39	5128	5148	5169	5189	5210	5230	5251	5271	5292	5313
40	5333	5354	5375	5396	5417	5438	5458	5479	5500	5522
41	5543	5564	5585	5606	5627	5649	5670	5691	5713	5734
42	5756	5777	5799	5820	5842	5863	5885	5907	5929	5950
43	5972	5994	6016	6038	6060	6082	6104	6126	6148	6170
44	6193	6215	6237	6259	6282	6304	6327	6349	6372	6394
45	6417	6439	6462	6485	6507	6530	6553	6576	6599	6622
46	6644	6667	6690	6714	6737	6760	6783	6806	6829	6853
47	6876	6899	6923	6946	6970	6993	7017	7040	7064	7087
48	7111	7135	7159	7182	7206	7230	7254	7278	7302	7326
49	7350	7374	7398	7422	7447	7471	7495	7519	7544	7568
50	7593	7617	7642	7666	7691	7715	7740	7765	7789	7814
51	7839	7864	7889	7914	7938	7963	7988	8014	8039	8064
52	8089	8114	8139	8165	8190	8215	8241	8266	8292	8317
53	8343	8368	8394	8419	8445	8471	8497	8522	8548	8574
54	8600	8626	8652	8678	8704	8730	8756	8782	8809	8835
55	8861	8887	8914	8940	8967	8993	9020	9046	9073	9099
56	9126	9153	9179	9206	9233	9260	9287	9314	9340	9367
57	9394	9422	9449	9476	9503	9530	9557	9585	9612	9639
58	9667	9694	9722	9749	9777	9804	9832	9859	9887	9915
59	9943	9970	9998	10026	10054	10082	10110	10138	10166	10194
60	10222	10250	10279	10307	10335	10363	10392	10420	10449	10477
	.0	.1	.2	.3	.4	.5	.6	.7	.8	.9

TABLE No. 10.—LEVEL CUTTINGS. $\frac{s+s'}{2} = \frac{1}{2}$; $b = 28\,feet.$

F.	.0	.1	.2	.3	.4	.5	.6	.7	.8	.9
0	0.	10.4	20.8	31.3	41.8	52.3	62.9	73.5	84.1	94.8
1	105.6	116.3	127.1	137.9	148.8	159.7	170.7	181.6	192.7	203.7
2	214.8	225.9	237.1	248.3	259.6	270.8	282.1	293.5	304.9	316.3
3	327.8	339.3	350.8	362.4	374.0	385.6	397.3	409.1	420.8	432.6
4	444.4	456.3	468.2	480.2	492.1	504.2	516.2	528.3	540.4	552.6
5	564.8	577.1	589.3	601.6	614.0	626.4	638.8	651.3	663.8	676.3
6	688.9	701.5	714.1	726 8	739.6	752.3	765.1	777.9	790.8	803.7
7	816.7	829.6	842.7	855.7	868.8	881.9	895.1	908.3	921.6	934.8
8	948.1	961.5	974.9	988.3	1002	1015	1029	1042	1056	1070
9	1083	1097	1111	1125	1138	1152	1166	1180	1194	1208
10	1222	1236	1250	1265	1279	1293	1307	1322	1336	1350
11	1365	1379	1394	1408	1423	1438	1452	1467	1482	1496
12	1511	1526	1541	1556	1571	1586	1601	1616	1631	1646
13	1661	1676	1692	1707	1722	1738	1753	1768	1784	1799
14	1815	1830	1846	1862	1877	1893	1909	1925	1940	1956
15	1972	1988	2004	2020	2036	2052	2068	2085	2101	2117
16	2133	2150	2166	2182	2199	2215	2232	2246	2265	2282
17	2298	2315	2332	2348	2365	2382	2399	2416	2433	2450
18	2467	2484	2501	2518	2535	2552	2570	2587	2604	2622
19	2639	2656	2674	2691	2709	2726	2744	2762	2779	2797
20	2815	2833	2850	2868	2886	2904	2922	2940	2958	2976
21	2994	3013	3031	3049	3067	3086	3104	3122	3141	3159
22	3178	3196	3215	3234	3252	3271	3290	3308	3327	3346
23	3365	3384	3403	3422	3441	3460	3479	3498	3517	3536
24	3556	3575	3594	3614	3633	3652	3672	3691	3711	3730
25	3750	3770	3789	3809	3829	3849	3868	3888	3908	3928
26	3946	3968	3988	4008	4028	4049	4069	4089	4109	4130
27	4150	4170	4191	4211	4232	4252	4273	4294	4314	4335
28	4356	4376	4397	4418	4439	4460	4481	4502	4523	4544
29	4565	4586	4607	4628	4650	4671	4692	4714	4735	4756
30	4778	4799	4821	4842	4864	4886	4907	4929	4951	4973
31	4994	5016	5038	5060	5082	5104	5126	5148	5170	5193
32	5215	5237	5259	5282	5304	5326	5349	5371	5394	5416
33	5439	5462	5484	5507	5530	5552	5575	5598	5621	5644
34	5667	5690	5713	5736	5759	5782	5805	5828	5852	5875
35	5898	5922	5945	5968	5992	6015	6039	6062	6086	6110
36	6133	6157	6181	6205	6228	6252	6276	6300	6324	6348
37	6372	6396	6420	6445	6469	6493	6517	6542	6566	6590
38	6615	6639	6664	6688	6713	6738	6762	6787	6812	6836
39	6861	6886	6911	6936	6961	6986	7011	7036	7061	7086
40	7111	7136	7162	7187	7212	7238	7263	7288	7314	7339
41	7365	7390	7416	7442	7467	7493	7519	7545	7570	7596
42	7622	7648	7674	7700	7726	7752	7778	7805	7831	7857
43	7883	7910	7936	7962	7989	8015	8042	8068	8095	8122
44	8148	8175	8202	8228	8255	8282	8309	8336	8363	8390
45	8417	8444	8471	8498	8525	8552	8580	8607	8634	8662
46	8689	8716	8744	8771	8799	8826	8854	8882	8909	8937
47	8965	8993	9020	9048	9076	9104	9132	9160	9188	9216
48	9244	9273	9301	9329	9357	9386	9414	9442	9471	9499
49	9528	9556	9585	9614	9642	9671	9700	9728	9757	9786
50	9815	9844	9873	9902	9931	9960	9989	10018	10047	10076
51	10106	10135	10164	10194	10223	10252	10282	10311	10341	10370
52	10400	10430	10459	10489	10519	10549	10578	10608	10638	10668
53	10698	10728	10758	10788	10818	10849	10879	10909	10939	10970
54	11000	11030	11061	11091	11122	11152	11183	11214	11244	11275
55	11306	11336	11367	11398	11429	11460	11491	11522	11553	11584
56	11615	11646	11677	11708	11740	11771	11802	11834	11865	11896
57	11928	11959	11991	12022	12054	12086	12117	12149	12181	12213
58	12244	12276	12308	12340	12372	12404	12436	12468	12500	12533
59	12565	12597	12629	12662	12694	12726	12759	12791	12824	12856
60	12889	12922	12954	12987	13020	13052	13085	13118	13151	13184
	.0	.1	.2	.3	.4	.5	.6	.7	.8	.9

TABLE No. 11.

Plus Corrections for $\frac{s+s'}{2} = \frac{1}{2}$.

Feet.	.0	.1	.2	.3	.4	.5	.6	.7	.8	.9
0	0.	0.	0.	0.	0.	0.	0.1	0.1	0.1	0.1
1	0.2	0.2	0.2	0.3	0.3	0.3	0.4	0.4	0.5	0.6
2	0.6	0.7	0.7	0.8	0.9	1.0	1.0	1.1	1.2	1.3
3	1.4	1.5	1.6	1.7	1.8	1.9	2.0	2.1	2.2	2.3
4	2.5	2.6	2.7	2.9	3.0	3.1	3.3	3.4	3.6	3.7
5	3.9	4.0	4.2	4.3	4.5	4.7	4.8	5.0	5.2	5.4
6	5.6	5.7	5.9	6.1	6.3	6.5	6.7	6.9	7.1	7.3
7	7.6	7.8	8.0	8.2	8.5	8.7	8.9	9.1	9.4	9.6
8	9.9	10.1	10.4	10.6	10.9	11.1	11.4	11.7	12.0	12.2
9	12.5	12.8	13.1	13.3	13.6	13.9	14.2	14.5	14.8	15.1
10	15.4	15.7	16.1	16.4	16.7	17.0	17.3	17.7	18.0	18.3
11	18.7	19.0	19.4	19.7	20.1	20.4	20.8	21.1	21.5	21.9
12	22.2	22.6	23.0	23.3	23.7	24.1	24.5	24.9	25.3	25.7
13	26.1	26.5	26.9	27.3	27.7	28.1	28.5	29.0	29.4	29.8
14	30.2	30.7	31.1	31.6	32.0	32.4	32.9	33.3	33.8	34.3
15	34.7	35.2	35.7	36.1	36.6	37.1	37.6	38.0	38.5	39.0
16	39.5	40.0	40.5	41.0	41.5	42.0	42.5	43.0	43.6	44.1
17	44.6	45.1	45.7	46.2	46.7	47.3	47.8	48.3	48.9	49.4
18	50.0	50.6	51.1	51.7	52.2	52.8	53.4	54.0	54.5	55.1
19	55.7	56.3	56.9	57.5	58.1	58.7	59.3	59.9	60.5	61.1
20	61.7	62.3	63.0	63.6	64.2	64.9	65.5	66.1	66.8	67.4
21	68.1	68.7	69.4	70.0	70.7	71.3	72.0	72.7	73.3	74.0
22	74.7	75.4	76.1	76.7	77.4	78.1	78.8	79.5	80.2	80.9
23	81.6	82.3	83.1	83.8	84.5	85.2	86.0	86.7	87.4	88.1
24	88.9	89.6	90.4	91.1	91.9	92.6	93.4	94.1	94.9	95.7
25	96.5	97.2	98.0	98.8	99.6	100.3	101.1	101.9	102.7	103.5
26	104.3	105.1	105.9	106.7	107.6	108.4	109.2	110.0	110.8	111.7
27	112.5	113.3	114.2	115.0	115.9	116.7	117.6	118.4	119.3	120.1
28	121.0	121.9	122.7	123.6	124.5	125.3	126.2	127.1	128.0	128.9
29	129.8	130.7	131.6	132.5	133.4	134.3	135.2	136.1	137.0	138.0
30	138.9	139.8	140.7	141.7	142.6	143.6	144.5	145.4	146.4	147.3
31	148.3	149.3	150.2	151.2	152.2	153.1	154.1	155.1	156.1	157.0
32	158.0	159.0	160.0	161.0	162.0	163.0	164.0	165.0	166.0	167.0
33	168.1	169.1	170.1	171.1	172.2	173.2	174.2	175.3	176.3	177.3
34	178.4	179.4	180.5	181.6	182.6	183.7	184.7	185.8	186.9	188.0
35	189.0	190.1	191.2	192.3	193.4	194.5	195.6	196.7	197.8	198.9
36	200.0	201.1	202.2	203.3	204.5	205.6	206.7	207.9	209.0	210.1
37	211.3	212.4	213.6	214.7	215.9	217.0	218.2	219.3	220.5	221.7
38	222.8	224.0	225.2	226.4	227.6	228.7	229.9	231.1	232.3	233.5
39	234.7	235.9	237.1	238.3	239.6	240.8	242.0	243.2	244.5	245.7
40	246.9	248.1	249.4	250.6	251.9	253.1	254.4	255.6	256.9	258.1
	.0	.1	.2	.3	.4	.5	.6	.7	.8	.9

Minus Corrections for $\frac{s+s'}{2} = \frac{1}{4}$.

NOTE.—The quantities from the above table divided by two give the plus corrections for $\frac{s+s'}{2} = \frac{1}{4}$.

TABLE No. 12.—LEVEL CUTTINGS. $\frac{s+s'}{2}=1$; $b=18$ feet.

Ft.	.0	.1	.2	.3	.4	.5	.6	.7	.8	.9
0	0.0	6.7	13.5	20.3	27.3	34.3	41.3	48.5	55.7	63.0
1	70.4	77.8	85.3	92.9	100.6	108.3	116.1	124.0	132.0	140.0
2	148.1	156.3	164.6	172.9	181.3	189.8	198.4	207.0	215.7	224.5
3	233.3	242.3	251.3	260.3	269.5	278.7	288.0	297.4	306.8	316.3
4	325.9	335.6	345.3	355.1	365.0	375.0	385.0	395.1	405.3	415.6
5	425.9	436.3	446.8	457.4	468.0	478.7	489.5	500.3	511.3	522.3
6	533.3	544.5	555.7	567.0	578.4	589.8	601.3	612.9	624.6	636.3
7	648.1	660.0	672.0	684.0	696.1	708.3	720.6	732.9	745.3	757.8
8	770.4	783.0	795.7	808.5	821.3	834.3	847.3	860.3	873.5	886.7
9	900.0	913.4	926.8	940.3	953.9	967.6	981.3	995.1	1009	1023
10	1037	1051	1065	1080	1094	1108	1123	1137	1152	1167
11	1181	1196	1211	1226	1241	1256	1272	1287	1302	1318
12	1333	1349	1365	1380	1396	1412	1428	1444	1460	1476
13	1493	1509	1525	1542	1558	1575	1592	1608	1625	1642
14	1659	1676	1693	1711	1728	1745	1763	1780	1798	1816
15	1833	1851	1869	1887	1905	1923	1941	1960	1978	1996
16	2015	2033	2052	2071	2089	2108	2127	2146	2165	2184
17	2204	2223	2242	2262	2281	2301	2321	2340	2360	2380
18	2400	2420	2440	2460	2481	2501	2521	2542	2562	2583
19	2604	2624	2645	2666	2687	2708	2729	2751	2772	2793
20	2815	2836	2858	2880	2901	2923	2945	2967	2989	3011
21	3033	3056	3078	3100	3123	3145	3168	3191	3213	3236
22	3259	3282	3305	3328	3352	3375	3398	3422	3445	3469
23	3493	3516	3540	3564	3588	3612	3636	3660	3685	3709
24	3733	3758	3782	3807	3832	3856	3881	3906	3931	3956
25	3981	4007	4032	4057	4083	4108	4134	4160	4185	4211
26	4237	4263	4289	4315	4341	4368	4394	4420	4447	4473
27	4500	4527	4553	4580	4607	4634	4661	4688	4716	4743
28	4770	4798	4825	4853	4881	4908	4936	4964	4992	5020
29	5048	5076	5105	5133	5161	5190	5218	5247	5276	5304
30	5333	5362	5391	5420	5449	5479	5508	5537	5567	5596
31	5626	5656	5685	5715	5745	5775	5805	5835	5865	5896
32	5926	5956	5987	6017	6048	6079	6109	6140	6171	6202
33	6233	6264	6296	6327	6358	6390	6421	6453	6485	6516
34	6548	6580	6612	6644	6676	6708	6741	6773	6805	6838
35	6870	6903	6936	6968	7001	7034	7067	7100	7133	7167
36	7200	7233	7267	7300	7334	7368	7401	7435	7469	7503
37	7537	7571	7605	7640	7674	7708	7743	7777	7812	7847
38	7881	7916	7951	7986	8021	8056	8092	8127	8162	8198
39	8233	8269	8305	8340	8376	8412	8448	8484	8520	8556
40	8593	8629	8665	8702	8738	8775	8812	8848	8885	8922
41	8959	8996	9033	9071	9108	9145	9183	9220	9258	9296
42	9333	9371	9409	9447	9485	9523	9561	9600	9638	9676
43	9715	9753	9792	9831	9869	9908	9947	9986	10025	10064
44	10104	10143	10182	10222	10261	10301	10341	10380	10420	10460
45	10500	10540	10580	10620	10661	10701	10741	10782	10822	10863
46	10904	10944	10985	11026	11067	11108	11149	11191	11232	11273
47	11315	11356	11398	11440	11481	11523	11565	11607	11649	11691
48	11733	11776	11818	11860	11903	11945	11988	12031	12073	12116
49	12159	12202	12245	12288	12332	12375	12418	12462	12505	12549
50	12593	12636	12680	12724	12768	12812	12856	12900	12945	12989
51	13033	13078	13122	13167	13212	13256	13301	13346	13391	13436
52	13481	13527	13572	13617	13663	13708	13754	13800	13845	13891
53	13937	13983	14029	14075	14121	14168	14214	14260	14307	14353
54	14400	14447	14493	14540	14587	14634	14681	14728	14776	14823
55	14870	14918	14965	15013	15061	15108	15156	15204	15252	15300
56	15348	15396	15445	15493	15541	15590	15638	15687	15736	15784
57	15833	15882	15931	15980	16029	16079	16128	16177	16227	16276
58	16326	16376	16425	16475	16525	16575	16625	16675	16725	16776
59	16826	16876	16927	16977	17028	17079	17129	17180	17231	17282
60	17333	17384	17436	17487	17538	17590	17641	17693	17745	17796
	.0	.1	.2	.3	.4	.5	.6	.7	.8	.9

TABLE No. 13.—LEVEL CUTTINGS. $\frac{s+s'}{2}=1$; $b=30$ feet.

F.	.0	.1	.2	.3	.4	.5	.6	.7	.8	.9
0	00.0	11.1	22.4	33.7	45.0	56.5	68.0	79.6	91.3	103.0
1	114.8	126.7	138.7	150.7	162.8	175.0	187.3	199.6	212.0	224.5
2	237.0	249.7	262.4	275.1	288.0	300.9	313.9	327.0	340.1	353.4
3	366.7	380.0	393.5	407.0	420.6	434.3	448.0	461.8	475.7	489.7
4	503.7	517.8	532.0	546.3	560.6	575.0	589.5	604.0	618.7	633.4
5	648.1	663.0	677.9	692.9	708.0	723.1	738.4	753.7	769.0	784.5
6	800.0	815.6	831.3	847.0	862.8	878.7	894.7	910.7	926.8	943.0
7	959.3	975.6	992.0	1008	1025	1042	1058	1075	1092	1109
8	1126	1143	1160	1177	1195	1212	1229	1247	1265	1282
9	1300	1318	1336	1354	1372	1390	1408	1426	1445	1463
10	1481	1500	1519	1537	1556	1575	1594	1613	1632	1651
11	1670	1690	1709	1728	1748	1768	1787	1807	1827	1847
12	1867	1887	1907	1927	1947	1968	1988	2008	2029	2050
13	2070	2091	2112	2133	2154	2175	2196	2217	2239	2260
14	2281	2303	2325	2346	2368	2390	2412	2434	2456	2478
15	2500	2522	2545	2567	2589	2612	2635	2657	2680	2703
16	2726	2749	2772	2795	2818	2842	2865	2888	2912	2936
17	2959	2983	3007	3031	3055	3079	3103	3127	3151	3176
18	3200	3224	3249	3274	3298	3323	3348	3373	3398	3423
19	3448	3473	3499	3524	3549	3575	3601	3626	3652	3678
20	3704	3730	3756	3782	3808	3834	3861	3887	3913	3940
21	3967	3993	4020	4047	4074	4101	4128	4155	4182	4210
22	4237	4264	4292	4320	4347	4375	4403	4431	4459	4487
23	4515	4543	4571	4600	4628	4656	4685	4714	4742	4771
24	4800	4829	4858	4887	4916	4945	4975	5004	5033	5063
25	5093	5122	5152	5182	5212	5242	5272	5302	5332	5362
26	5393	5423	5453	5484	5515	5545	5576	5607	5638	5669
27	5700	5731	5762	5794	5825	5856	5888	5920	5951	5983
28	6015	6047	6079	6111	6143	6175	6207	6240	6272	6304
29	6337	6370	6402	6435	6468	6501	6534	6567	6600	6633
30	6667	6700	6733	6767	6801	6834	6868	6902	6936	6970
31	7004	7038	7072	7106	7141	7175	7209	7244	7279	7313
32	7348	7383	7418	7453	7488	7523	7558	7594	7629	7664
33	7700	7736	7771	7807	7843	7879	7915	7951	7987	8023
34	8059	8096	8132	8168	8205	8242	8278	8315	8352	8389
35	8426	8463	8500	8537	8575	8612	8649	8687	8725	8762
36	8800	8838	8876	8914	8952	8990	9028	9066	9105	9143
37	9181	9220	9259	9297	9336	9375	9414	9453	9492	9531
38	9570	9610	9649	9688	9728	9768	9807	9847	9887	9927
39	9967	10007	10047	10087	10127	10168	10208	10248	10289	10330
40	10370	10411	10452	10493	10534	10575	10616	10657	10699	10740
41	10781	10823	10865	10906	10948	10990	11032	11074	11116	11158
42	11200	11242	11285	11327	11369	11412	11455	11497	11540	11583
43	11626	11669	11712	11755	11798	11842	11885	11928	11972	12016
44	12059	12103	12147	12191	12235	12279	12323	12367	12411	12456
45	12500	12544	12589	12634	12678	12723	12768	12813	12858	12903
46	12948	12993	13039	13084	13129	13175	13221	13266	13312	13358
47	13404	13450	13496	13542	13588	13634	13681	13727	13773	13820
48	13867	13913	13960	14007	14054	14101	14148	14195	14242	14290
49	14337	14384	14432	14480	14527	14575	14623	14671	14719	14767
50	14815	14863	14911	14960	15008	15056	15105	15154	15202	15251
51	15300	15349	15398	15447	15496	15545	15595	15644	15693	15743
52	15793	15842	15892	15942	15992	16042	16092	16142	16192	16242
53	16293	16343	16393	16444	16495	16545	16596	16647	16698	16749
54	16800	16851	16902	16954	17005	17056	17108	17160	17211	17263
55	17315	17367	17419	17471	17523	17575	17627	17680	17732	17784
56	17837	17890	17942	17995	18048	18101	18154	18207	18260	18313
57	18367	18420	18473	18527	18581	18634	18688	18742	18796	18850
58	18904	18958	19012	19066	19121	19175	19229	19284	19339	19393
59	19448	19503	19558	19613	19668	19723	19778	19834	19889	19944
60	20000	20056	20111	20167	20223	20279	20335	20391	20447	20503
	.0	.1	.2	.3	.4	.5	.6	.7	.8	.9

TABLE No. 14.

Plus Corrections for $\frac{s+s'}{2} = 1$.

Feet.	.0	.1	.2	.3	.4	.5	.6	.7	.8	.9
0	0.0	0.0	0.0	0.0	0.0	0.1	0.1	0.2	0.2	0.3
1	0.3	0.4	0.4	0.5	0.6	0.7	0.8	0.9	1.0	1.1
2	1.2	1.4	1.5	1.6	1.8	1.9	2.1	2.2	2.4	2.6
3	2.8	3.0	3.2	3.4	3.6	3.8	4.0	4.2	4.5	4.7
4	4.9	5.2	5.4	5.7	6.0	6.3	6.5	6.8	7.1	7.4
5	7.7	8.0	8.3	8.7	9.0	9.3	9.7	10.0	10.4	10.7
6	11.1	11.5	11.9	12.3	12.6	13.0	13.4	13.9	14.3	14.7
7	15.1	15.6	16.0	16.4	16.9	17.4	17.8	18.3	18.8	19.3
8	19.8	20.3	20.8	21.3	21.8	22.3	22.8	23.4	23.9	24.4
9	25.0	25.6	26.1	26.7	27.3	27.9	28.4	29.0	29.6	30.3
10	30.9	31.5	32.1	32.7	33.4	34.0	34.7	35.3	36.0	36.7
11	37.3	38.0	38.7	39.4	40.1	40.8	41.5	42.3	43.0	43.7
12	44.4	45.2	45.9	46.7	47.5	48.2	49.0	49.8	50.6	51.4
13	52.2	53.0	53.8	54.6	55.4	56.2	57.1	57.9	58.8	59.6
14	60.5	61.4	62.2	63.1	64.0	64.9	65.8	66.7	67.6	68.5
15	69.4	70.4	71.3	72.3	73.2	74.2	75.1	76.1	77.0	78.0
16	79.0	80.0	81.0	82.0	83.0	84.0	85.0	86.1	87.1	88.2
17	89.2	90.3	91.3	92.4	93.4	94.5	95.6	96.7	97.8	98.9
18	100.0	101.1	102.2	103.4	104.5	105.6	106.8	107.9	109.1	110.2
19	111.4	112.6	113.8	115.0	116.2	117.4	118.6	119.8	121.0	122.2
20	123.5	124.7	125.9	127.2	128.4	129.7	131.0	132.3	133.5	134.8
21	136.1	137.4	138.7	140.0	141.3	142.7	144.0	145.3	146.7	148.0
22	149.4	150.7	152.1	153.5	154.9	156.3	157.6	159.0	160.4	161.9
23	163.3	164.7	166.1	167.6	169.0	170.4	171.9	173.4	174.8	176.3
24	177.8	179.3	180.8	182.3	183.8	185.3	186.8	188.3	189.8	191.4
25	192.9	194.4	196.0	197.6	199.1	200.7	202.3	203.9	205.4	207.0
26	208.6	210.3	211.9	213.5	215.1	216.7	218.4	220.0	221.7	223.3
27	225.0	226.7	228.3	230.0	231.7	233.4	235.1	236.8	238.5	240.3
28	242.0	243.7	245.4	247.2	248.9	250.7	252.5	254.2	256.0	257.8
29	259.6	261.4	263.2	265.0	266.8	268.6	270.4	272.2	274.1	275.9
30	277.8	279.6	281.5	283.4	285.2	287.1	289.0	290.9	292.8	294.7
31	296.6	298.5	300.4	302.4	304.3	306.3	308.2	310.2	312.1	314.1
32	316.0	318.0	320.0	322.0	324.0	326.0	328.0	330.0	332.0	334.1
33	336.1	338.2	340.2	342.3	344.3	346.4	348.4	350.5	352.6	354.7
34	356.8	358.9	361.0	363.1	365.2	367.4	369.5	371.6	373.8	375.9
35	378.1	380.2	382.4	384.6	386.8	389.0	391.2	393.4	395.6	397.8
36	400.0	402.2	404.5	406.7	408.9	411.2	413.4	415.7	418.0	420.3
37	422.5	424.8	427.1	429.4	431.7	434.0	436.3	438.7	441.0	443.3
38	445.7	448.0	450.4	452.7	455.1	457.5	459.9	462.3	464.6	467.0
39	469.4	471.9	474.3	476.7	479.1	481.6	484.0	486.4	488.9	491.4
40	493.8	496.3	498.8	501.3	503.8	506.2	508.8	511.3	513.8	516.3
	.0	.1	.2	.3	.4	.5	.6	.7	.8	.9

Minus Corrections for $\frac{s+s'}{2} = \frac{1}{2}$.

NOTE.—For minus corrections for $\frac{s+s'}{2} = 1$, see Table 5.

TABLE No. 15.—Level Cuttings. $\frac{s+s'}{2}=1\frac{1}{2}$; $b = 14\ feet$.

h.	.0	.1	.2	.3	.4	.5	.6	.7	.8	.9
0	0.0	5.2	10.6	16.1	21.6	27.3	33.1	39.0	45.0	51.2
1	57.4	63.8	70.2	76.8	83.5	90.3	97.2	104.2	111.3	118.6
2	125.9	133.4	141.0	148.6	156.4	164.4	172.4	180.5	188.7	197.1
3	205.6	214.1	222.8	231.6	240.5	249.5	258.7	267.9	277.3	286.7
4	296.3	306.0	315.8	325.7	335.7	345.8	356.1	366.4	376.9	387.5
5	398.1	408.9	419.9	430.9	442.0	453.2	464.6	476.1	487.6	499.3
6	511.1	523.0	535.0	547.2	559.4	571.8	584.2	596.8	609.5	622.3
7	635.2	648.2	661.3	674.6	687.9	701.4	715.0	728.6	742.4	756.4
8	770.4	784.5	798.7	813.1	827.6	842.1	856.8	871.6	886.5	901.5
9	916.7	931.9	947.3	962.7	978.3	994.0	1010	1026	1042	1058
10	1074	1090	1107	1123	1140	1157	1174	1191	1208	1225
11	1243	1260	1278	1295	1313	1331	1349	1367	1385	1404
12	1422	1441	1459	1478	1497	1516	1535	1555	1574	1593
13	1613	1633	1652	1672	1692	1713	1733	1753	1774	1794
14	1815	1836	1857	1878	1899	1920	1941	1963	1984	2006
15	2028	2050	2072	2094	2116	2138	2161	2183	2206	2229
16	2252	2275	2298	2321	2345	2368	2392	2415	2439	2463
17	2487	2511	2535	2560	2584	2609	2633	2658	2683	2708
18	2733	2759	2784	2809	2835	2861	2886	2912	2938	2965
19	2991	3017	3044	3070	3097	3124	3151	3178	3205	3232
20	3259	3287	3314	3342	3370	3398	3426	3454	3482	3510
21	3539	3567	3596	3625	3654	3683	3712	3741	3771	3800
22	3830	3859	3889	3919	3949	3979	4009	4040	4070	4101
23	4131	4162	4193	4224	4255	4287	4318	4349	4381	4413
24	4444	4476	4508	4541	4573	4605	4638	4670	4703	4736
25	4769	4802	4835	4868	4901	4935	4968	5002	5036	5070
26	5104	5138	5172	5206	5241	5275	5310	5345	5380	5415
27	5450	5485	5521	5556	5592	5627	5663	5699	5735	5771
28	5807	5844	5880	5917	5953	5990	6027	6064	6101	6139
29	6176	6213	6251	6289	6326	6364	6402	6441	6479	6517
30	6556	6594	6633	6672	6711	6750	6789	6828	6867	6907
31	6946	6986	7026	7066	7106	7146	7186	7226	7267	7307
32	7348	7389	7430	7471	7512	7553	7595	7636	7678	7719
33	7761	7803	7845	7887	7929	7972	8014	8057	8099	8142
34	8185	8228	8271	8315	8358	8401	8445	8489	8532	8576
35	8620	8665	8709	8753	8798	8842	8887	8932	8977	9022
36	9067	9112	9157	9203	9248	9294	9340	9386	9432	9478
37	9524	9570	9617	9663	9710	9757	9804	9851	9898	9945
38	9993	10040	10088	10135	10183	10231	10279	10327	10375	10424
39	10472	10521	10569	10618	10667	10716	10765	10815	10864	10913
40	10963	11013	11062	11112	11162	11213	11263	11313	11364	11414
41	11465	11516	11567	11618	11669	11720	11771	11823	11874	11926
42	11978	12030	12082	12134	12186	12238	12291	12343	12396	12449
43	12502	12555	12608	12661	12715	12768	12822	12875	12929	12983
44	13037	13091	13145	13200	13254	13309	13363	13418	13473	13528
45	13583	13639	13694	13749	13805	13861	13916	13972	14028	14085
46	14141	14197	14254	14310	14367	14424	14481	14538	14595	14652
47	14709	14767	14824	14882	14940	14998	15056	15114	15172	15230
48	15289	15347	15406	15465	15524	15583	15642	15701	15761	15820
49	15880	15939	15999	16059	16119	16179	16239	16300	16360	16421
50	16481	16542	16603	16664	16725	16787	16848	16909	16971	17033
51	17094	17156	17218	17281	17343	17405	17468	17530	17593	17656
52	17719	17782	17845	17908	17971	18035	18098	18162	18226	18290
53	18354	18418	18482	18546	18611	18675	18740	18805	18870	18935
54	19000	19065	19131	19196	19262	19327	19393	19459	19525	19591
55	19657	19724	19790	19857	19923	19990	20057	20124	20191	20259
56	20326	20393	20461	20529	20596	20664	20732	20801	20869	20937
57	21006	21074	21143	21212	21281	21350	21419	21488	21557	21627
58	21696	21766	21836	21906	21976	22046	22116	22186	22257	22327
59	22398	22469	22540	22611	22682	22753	22825	22896	22968	23039
60	23111	23183	23255	23327	23399	23472	23544	23617	23689	23762
	.0	.1	.2	.3	.4	.5	.6	.7	.8	.9

TABLE No. 16.—LEVEL CUTTINGS. $\frac{s+s}{2} = 1\frac{1}{2}$; $b = 26\,feet$.

	.0	.1	.2	.3	.4	.5	.6	.7	.8	.9
0	0.0	9.7	19.5	29.4	39.4	49.5	59.8	70.1	80.6	91.2
1	101.9	112.6	123.6	134.6	145.7	156.9	168.3	179.8	191.3	203.0
2	214.8	226.7	238.7	250.9	263.1	275.5	287.9	300.5	313.2	326.0
3	338.9	351.9	365.0	378.3	391.6	405.1	418.7	432.4	446.1	460.1
4	474.1	488.2	502.4	516.8	531.3	545.8	560.5	575.3	590.2	605.2
5	620.4	635.6	651.0	666.4	682.0	697.7	713.5	729.4	745.4	761.5
6	777.8	794.1	810.6	827.2	843.9	860.6	877.6	894.6	911.7	928.9
7	946.3	963.8	981.3	999.0	1017	1035	1053	1071	1089	1107
8	1126	1145	1163	1182	1201	1220	1239	1258	1278	1297
9	1317	1336	1356	1376	1396	1416	1436	1457	1477	1498
10	1519	1539	1560	1581	1602	1624	1645	1666	1688	1710
11	1732	1753	1775	1798	1820	1842	1865	1887	1910	1933
12	1956	1979	2002	2025	2048	2072	2095	2119	2143	2167
13	2191	2215	2239	2264	2288	2312	2337	2362	2387	2412
14	2437	2462	2488	2513	2539	2564	2590	2616	2642	2668
15	2694	2721	2747	2774	2801	2827	2854	2881	2908	2936
16	2963	2990	3018	3046	3074	3101	3129	3158	3186	3214
17	3243	3271	3300	3329	3358	3387	3416	3445	3474	3504
18	3533	3563	3593	3623	3653	3683	3713	3744	3774	3804
19	3835	3866	3897	3928	3959	3990	4022	4053	4085	4116
20	4148	4180	4212	4244	4276	4309	4341	4374	4407	4439
21	4472	4505	4538	4572	4605	4638	4672	4706	4740	4773
22	4807	4842	4876	4910	4945	4979	5014	5049	5084	5119
23	5154	5189	5224	5260	5295	5331	5367	5403	5439	5475
24	5511	5548	5584	5620	5657	5694	5731	5768	5805	5842
25	5880	5917	5955	5992	6030	6068	6106	6144	6182	6221
26	6259	6298	6337	6375	6414	6453	6492	6532	6571	6610
27	6650	6690	6730	6769	6809	6850	6890	6930	6971	7011
28	7052	7093	7134	7175	7216	7257	7298	7340	7381	7423
29	7465	7507	7549	7591	7633	7676	7718	7760	7803	7846
30	7889	7932	7975	8018	8062	8105	8149	8192	8236	8280
31	8324	8368	8412	8457	8501	8546	8591	8635	8680	8725
32	8770	8816	8861	8906	8952	8998	9044	9089	9135	9182
33	9228	9274	9321	9367	9414	9461	9508	9555	9602	9649
34	9696	9744	9791	9839	9887	9935	9983	10031	10079	10128
35	10176	10224	10273	10322	10371	10420	10469	10518	10568	10617
36	10667	10716	10766	10816	10866	10916	10966	11017	11067	11118
37	11169	11219	11270	11321	11372	11424	11475	11526	11578	11630
38	11682	11733	11785	11838	11890	11942	11995	12047	12100	12153
39	12206	12259	12312	12365	12418	12472	12525	12579	12633	12687
40	12741	12795	12849	12904	12958	13012	13067	13122	13177	13232
41	13287	13342	13398	13453	13509	13564	13620	13676	13732	13788
42	13844	13901	13957	14014	14071	14127	14184	14241	14298	14356
43	14413	14470	14528	14586	14644	14701	14759	14818	14876	14934
44	14993	15051	15110	15169	15228	15287	15346	15405	15464	15524
45	15583	15643	15703	15763	15823	15883	15943	16004	16064	16124
46	16185	16246	16307	16368	16429	16490	16552	16613	16675	16736
47	16798	16860	16922	16984	17046	17109	17171	17234	17297	17359
48	17422	17485	17548	17612	17675	17738	17802	17866	17930	17993
49	18057	18122	18186	18250	18315	18379	18444	18509	18574	18639
50	18704	18769	18834	18900	18965	19031	19097	19163	19229	19295
51	19361	19428	19494	19560	19627	19694	19761	19828	19895	19962
52	20030	20097	20165	20232	20300	20368	20436	20504	20572	20641
53	20709	20778	20847	20915	20984	21053	21122	21192	21261	21330
54	21400	21470	21540	21610	21679	21750	21820	21890	21961	22031
55	22102	22173	22244	22315	22386	22457	22528	22600	22671	22743
56	22815	22887	22959	23031	23103	23176	23248	23320	23393	23466
57	23539	23612	23685	23758	23832	23905	23979	24052	24126	24200
58	24274	24348	24422	24497	24571	24646	24721	24795	24870	24945
59	25020	25096	25171	25246	25322	25398	25474	25549	25625	25702
60	25778	25854	25931	26007	26084	26161	26238	26315	26392	26469
	.0	.1	.2	.3	.4	.5	.6	.7	.8	.9

TABLE No. 17.

Plus Corrections for $\frac{s+s'}{2} = 1\frac{1}{4}$.

Feet.	.0	.1	.2	.3	.4	.5	.6	.7	.8	.9
0	0.0	0.0	0.0	0.0	0.1	0.1	0.2	0.2	0.3	0.4
1	0.5	0.6	0.7	0.8	0.9	1.0	1.2	1.3	1.5	1.7
2	1.9	2.0	2.2	2.4	2.7	2.9	3.1	3.4	3.6	3.9
3	4.2	4.4	4.7	5.0	5.4	5.7	6.0	6.3	6.7	7.0
4	7.4	7.8	8.2	8.6	9.0	9.4	9.8	10.2	10.7	11.1
5	11.6	12.0	12.5	13.0	13.5	14.0	14.5	15.0	15.6	16.1
6	16.7	17.2	17.8	18.4	19.0	19.6	20.2	20.8	21.4	22.0
7	22.7	23.3	24.0	24.7	25.4	26.0	26.7	27.4	28.2	28.9
8	29.6	30.4	31.1	31.9	32.7	33.4	34.2	35.0	35.9	36.7
9	37.5	38.3	39.2	40.0	40.9	41.8	42.7	43.6	44.5	45.4
10	46.3	47.2	48.2	49.1	50.1	51.0	52.	53.	54.	55.
11	56.	57.	58.1	59.1	60.2	61.2	62.3	63.4	64.5	65.6
12	66.7	67.8	68.9	70.	71.2	72.3	73.5	74.7	75.9	77.
13	78.2	79.4	80.7	81.9	83.1	84.4	85.6	86.9	88.2	89.4
14	90.7	92.0	93.4	94.7	96.0	97.3	98.7	100.	101.4	102.8
15	104.2	105.6	107.0	108.4	109.8	111.2	112.7	114.1	115.6	117.
16	118.5	120.	121.5	123.	124.5	126.	127.6	129.1	130.7	132.2
17	133.8	135.4	137.0	138.6	140.2	141.8	143.4	145.	146.7	148.3
18	150.	151.7	153.4	155.	156.7	158.4	160.2	161.9	163.6	165.4
19	167.1	168.9	170.7	172.4	174.2	176.0	177.9	179.7	181.5	183.3
20	185.2	187.	188.9	190.8	192.7	194.6	196.5	198.4	200.3	202.2
21	204.2	206.1	208.1	210.	212.	214.	216.	218.	220.	222.
22	224.1	226.1	228.2	230.2	232.3	234.4	236.5	238.6	240.7	242.8
23	244.9	247.	249.2	251.3	253.5	255.7	257.9	260.0	262.2	264.4
24	266.7	268.9	271.1	273.4	275.6	277.9	280.2	282.4	284.7	287.0
25	289.4	291.7	294.	296.3	298.7	301.0	303.4	305.8	308.2	310.6
26	313.	315.4	317.8	320.2	322.7	325.1	327.6	330.0	332.5	335.
27	337.5	340.0	342.5	345.0	347.6	350.1	352.7	355.2	357.8	360.4
28	363.0	365.6	368.2	370.8	373.4	376.0	378.7	381.3	384.0	386.7
29	389.4	392.0	394.7	397.4	400.2	402.9	405.6	408.4	411.1	413.9
30	416.7	419.4	422.2	425.0	427.9	430.7	433.5	436.3	439.2	442.0
31	444.9	447.8	450.7	453.6	456.5	459.4	462.3	465.2	468.2	471.1
32	474.1	477.0	480.0	483.0	486.0	489.0	492.0	495.0	498.1	501.1
33	504.2	507.2	510.3	513.4	516.5	519.6	522.7	525.8	528.9	532.0
34	535.2	538.3	541.5	544.7	547.9	551.0	554.2	557.4	560.7	563.9
35	567.1	570.4	573.6	576.9	580.2	583.4	586.7	590.0	593.4	596.7
36	600.0	603.3	606.7	610.0	613.4	616.8	620.2	623.6	627.0	630.4
37	633.8	637.2	640.7	644.1	647.6	651.0	654.5	658.0	661.5	665.0
38	668.5	672.0	675.6	679.1	682.7	686.2	689.8	693.4	697.0	700.6
39	704.2	707.8	711.4	715.0	718.7	722.3	726.0	729.7	733.4	737.0
40	740.7	744.4	748.2	751.9	755.6	759.4	763.1	766.9	770.7	774.4
	.0	.1	.2	.3	.4	.5	.6	.7	.8	.9

Minus Corrections for $\frac{s+s'}{2} = \frac{3}{4}$.

NOTE.—The quantities from above table divided by two give the plus corrections for $\frac{s+s'}{2} = \frac{3}{4}$.

TABLE No. 18.

Factors for Correction of Contents on Curves.

$d\, f\, d''$ in feet.	Factor.	$d\, f\, d''$ in feet.	Factor.	$d\, f\, d''$ in feet.	Factor.	$d\, f\, d''$ in feet.	Factor.	$d\, f\, d''$ in feet.	Factor.
1	.00022	21	.00452	41	.00883	61	.01314	81	.01745
2	.00043	22	.00474	42	.00905	62	.01336	82	.01767
3	.00065	23	.00496	43	.00926	63	.01357	83	.01788
4	.00086	24	.00517	44	.00948	64	.01379	84	.01810
5	.00108	25	.00539	45	.00970	65	.01400	85	.01831
6	.00129	26	.00560	46	.00991	66	.01422	86	.01853
7	.00151	27	.00582	47	.01013	67	.01444	87	.01875
8	.00172	28	.00603	48	.01034	68	.01465	88	.01896
9	.00194	29	.00625	49	.01056	69	.01487	89	.01918
10	.00215	30	.00646	50	.01077	70	.01508	90	.01939
11	.00237	31	.00668	51	.01099	71	.01530	91	.01961
12	.00259	32	.00689	52	.01120	72	.01551	92	.01982
13	.00280	33	.00711	53	.01142	73	.01573	93	.02004
14	.00302	34	.00733	54	.01163	74	.01594	94	.02025
15	.00323	35	.00754	55	.01185	75	.01616	95	.02047
16	.00345	36	.00776	56	.01207	76	.01637	96	.02068
17	.00366	37	.00797	57	.01228	77	.01659	97	.02090
18	.00388	38	.00819	58	.01250	78	.01681	98	.02111
19	.00409	39	.00840	59	.01271	79	.01702	99	.02133
20	.00431	40	.00862	60	.01293	80	.01724	100	.02155

The Construction of Tables of Contents of Level Cuttings.

General Data.

Base $= b$; half sum of side slopes $= s$.

For each 0.1 of height, the second difference $= (0.074074+)s$.

Between heights 0.0 and 0.1 first difference $= \dfrac{10b+s}{27}$

" " 2.7 " 2.8 " " $= \dfrac{10b+s}{27} + 2s$

" " 5.4 " 5.5 " " $= \dfrac{10b+s}{27} + 4s$

Contents for a height of $0.1 = \dfrac{10b+s}{27}$

" " $2.7 = 10b + 27 \times s$

" " $5.4 = 20b + 27 \times 4s$

To write out a table of level cuttings progressing in height by tenths, rule five columns carried to heights of 2.7 when $s = 1$ or one of its multiples, and to heights of 5.4 when $s = \frac{1}{4}$ or one of its odd multiples.

Example.—(See portion of table given below) $b = 28$; $s = 1$. Here the second difference $= 0.074074+$; first difference between heights 0.0 and $0.1 = 10.407407+$; between 2.7 and $2.8 = 12.407407+$.

Place the heights from 0.0 to 2.8 in the first column; then put first difference 10.407407+ in third column opposite 0.0 in first, and second difference 0.074074+ immediately above the first difference.

As a test for the continued addition of the second difference, put the first difference 12.407407+ in its place in third column, opposite 2.7 in first. Now add 0.074074+ for each 0.1 of height up to 2.7, taking care to record the repeating fractions correctly, and see that the last addition gives 12.407407+ opposite 2.7. Then add each amount in third column to the amount on its left in second, recording each sum in the next line below, and keeping the repeating fractions correct. The contents in second column opposite 2.7 should be $= 10b + 27s = 307.0$.

Now repeat the amounts in the second column to the nearest tenth, placing them in the fourth column, and as before with regard to the heights in the first. From the fourth column, by subtraction, write the first differences anew, to the nearest tenth, in the fifth column, and opposite their respective positions in the third.

For the remainder of the table, rule columns in sets of threes; the first of each set to contain respectively the heights from 2.8 to 5.4, 5.5 to 8.1, 8.2 to 10.8, etc. Then increase each of the first differences in the 5th column by $2s = 2.0$, and the first differences from 2.8 to 5.4 are obtained for the eighth column. These again increased by 2.0 give the first differences from 5.5 to 8.1 for the eleventh column, etc. In this way the first differences for the whole table may be written to one place of decimals. Each first difference is to be added to the contents opposite in the next column on the left, and the sum recorded in the first line below. With contents calculated by Formula $C = (b+hs)$ $h \times \frac{100}{27}$ at intervals for tests, mistakes are almost impossible.

To carry out the table to whole numbers only, repeat the second column to the nearest whole number, get the first differences to whole numbers by subtraction, and proceed in all respects as above directed.*

(1)	(2)	(3)	(4)	(5)	(6)	(7)	(8)	(9)	(10)	(11)
Heights.	Contents.	0.074074	Contents.	1st Diff.	Heights.	Contents.	1st Diff.	Heights.	Contents.	1st Diff.
.0	0.000000	10.407407								
.1	10.407407	10.481481	10.4	10.5	2.8	319.4	12.5	5.5	682.4	14.5
.2	20.888888	10.555555	20.9	10.5	.9	331.9	12.5	.6	696.9	14.5
.3	31.444444	10.629629	31.4	10.7	3.0	344.4	12.7	.7	711.4	14.7
.4	42.074074	10.703703	42.1	10.7	.1	357.1	12.7	.8	726.1	14.7
.5	52.777777	10.777777	52.8	10.8	.2	369.8	12.8	.9	740.8	14.8
.6	63.555555	10.851851	63.6	10.8	.3	382.6	12.8	6.0	755.6	14.8
.7	74.407407	10.925925	74.4	10.9	.4	395.4	12.9	.1	770.4	14.9
.8	85.333333	11.0	85.3	11.0	.5	408.3	13.0	.2	785.3	15.0
.9	96.333333	11.074074	96.3	11.1	.6	421.3	13.1	.3	800.3	15.1
1.0	107.407407	11.148148	107.4	11.2	.7	434.4	13.2	.4	815.4	15.2
.1	118.555555	11.222222	118.6	11.2	.8	447.6	13.2	.5	830.6	15.2
.2	129.777777	11.296296	129.8	11.3	.9	460.8	13.3	.6	845.8	15.3
.3	141.074074	11.370370	141.1	11.3	4.0	474.1	13.3	.7	861.1	15.3
.4	152.444444	11.444444	152.4	11.5	.1	487.4	13.5	.8	876.4	15.5
.5	163.888888	11.518518	163.9	11.5	.2	500.9	13.5	.9	891.9	15.5
.6	175.407407	11.592592	175.4	11.6	.3	514.4	13.6	7.0	907.4	15.6
.7	187.0	11.666666	187.0	11.7	.4	528.0	13.7	.1	923.0	15.7
.8	198.666666	11.740740	198.7	11.7	.5	541.7	13.7	.2	938.7	15.7
.9	210.407407	11.814814	210.4	11.8	.6	555.4	13.8	.3		15.8
2.0	222.222222	11.888888	222.2	11.9	.7	569.2	13.9	.4		15.9
.1	234.111111	11.962962	234.1	12.0	.8	583.1	14.0	.5		16.0
.2	246.074074	12.037037	246.1	12.0	.9	597.1	14.0	.6		16.0
.3	258.111111	12.111111	258.1	12.1	5.0	611.1	14.1	.7		16.1
.4	270.222222	12.185185	270.2	12.2	.1	625.2	14.2	.8		16.2
.5	282.407407	12.259259	282.4	12.3	.2	639.4	14.3	.9		16.3
.6	294.666666	12.333333	294.7	12.3	.3	653.3	14.3	8.0		16.3
2.7	307.0	12.407407	307.0	12.4	.4	668.0	14.4	8.1	1083.0	16.4
2.8	319.407407		319.4							

* In case the second column does not give a whole number at the height of 2.7, it should be carried out to 5.4, or to the requisite multiple of 2.7.

www.ingramcontent.com/pod-product-compliance
Lightning Source LLC
Chambersburg PA
CBHW022149090426
42742CB00010B/1446